ELECTRICIAN'S EXAM

PREPARATION GUIDE

Sixth Edition

Based on the 2005 *NEC*®

by

John E. Traister

Revised and Updated by

Dale C. Brickner

Craftsman Book Company
6058 Corte del Cedro / P.O. Box 6500 / Carlsbad, CA 92018

Acknowledgments

I am indebted to several individuals and organizations who helped in the preparation of this book. One group is the electrical examining boards throughout the United States. A list of these organizations appears in Appendix I of this book. The following were especially helpful in furnishing reference materials or else helping with the production.

National Fire Protection Association, Inc. (NFPA)

C. Keeler Chapman, *Art Work*

Dionne L. Brickner, *Typist*

Nicole L. Brickner, *Organizer*

Joe A. Fintz, *Final Exam Verification*

Ron Murray, *Code Consultant*

Floyd Richards, *Question Verification & Code Consultant*

Library of Congress Cataloging-in-Publication Data

Traister, John E.
Electrician's exam preparation guide : based on the 2005 NEC / by John E. Traister ; revised and edited by Dale C. Brickner.-- 6th ed.
p. cm.
Includes index
ISBN 1-57218-152-4
ISBN-13: 978-1-57218-152-6
1. Electric engineering--Examinations, questions, etc. 2. Electric engineering--Examinations--Study guides. 3. Electricians--Licenses--United States. 4. National Fire Protection Association. National Electrical Code (2005) I. Brickner, Dale C. II. Title.
TK169.T73 2004
621.319'24'076--dc22 2004063446

Third printing 2007

Contents

Introduction — How to Use This Book

If you have been installing electrical systems for some time as an apprentice, helper, or unlicensed electrician, this book is for you. The information between the covers of this book will cover every subject that is likely to appear on most electrician's exams — either state or local

If you are just starting your career as an apprentice electrician, this book is also for you. It begins at the beginning. You will have no trouble understanding what is explained here. Read each page carefully and you will soon earn the recognition that licensed professionals are entitled to in our present society. The financial rewards are another factor which will make your efforts worthwhile.

In most communities, any electrician working *without supervision* must be licensed. For larger electrical construction projects, many states now require the certification of journeyman electricians as well as specialty electricians, such as splicers of high-voltage cable. This trend is certain to continue as legislatures recognize the need to protect the public from incompetents. The state of Virginia, for example, is now requiring all persons doing electrical work to be licensed.

Most licensing authorities prepare demanding exams that are a good test of the examinee's knowledge. These exams help to guarantee that electrical systems installed in building construction will meet minimum standards for protecting the lives and health of building occupants (and the buildings themselves) for many years to come. This also helps to keep insurance rates to a minimum.

Begin your study for any electrician's exam with two points in mind:

- Take the exam seriously
- Every minute spent studying this book increases your chances of passing the exam

You can pass any electrician's exam, but only if you study carefully each of the questions in this book. What you learn from studying is the foundation on which your professional career will be built.

Understand also that the licensing authority isn't the enemy. They aren't trying to keep you out of the electrical business. They only want to set some basic standards and be assured that your installations will be done in a workmanlike manner and in accordance with the latest edition of the National Electrical Code® *(NEC)*. The public should be assured that all licensed electricians are knowledgeable professionals. That's good for society in general, and it's good for all professional electricians and electrical contractors who live and work in your area.

Unfortunately, there are too many applicants who are not well prepared when they sit down to take the electrician's exam. Taking an electrician's exam without doing a good job of preparation is a complete waste of time — both yours and that of the licensing authority. The results are predictable. Don't make that mistake.

The most common reason for failure is that the applicant didn't study properly because he didn't know how, or studied the wrong material. This book should put an end to that excuse. You have in your hands the most complete, easiest-to-use, most practical reference available for preparing to take the tests that are actually given today. Read this book carefully, examine every question, understand all the answers. Do this, and there's no way you will be unprepared on examination day. You are almost certain to score high.

All the common questions and answers are here, but just knowing the answer is not always enough. Sometimes it is just as important to understand *why* a particular answer is correct. That's why

The National Electrical Code *(NEC)* has become the Bible of the electrical industry

many answers include a quotation or reference section from the National Electrical Code. Sometimes you will find notes or clarifications under the answer when there is an important point you might miss.

The National Electrical Code is used in practically every area of the United States for inspecting electrical systems in building construction. Most of the questions appearing on electrician's exams will come directly from Articles and Sections of the latest *NEC*. Therefore a brief review of the individual *NEC* sections that apply to electrical systems is in order. Sample questions concerning all sections of the *NEC* may be found in the chapters to follow.

This book, however, is not a substitute for the *NEC*. You need a copy of the most recent edition and it should be kept handy at all times. The more you know about the code, the more you are likely to refer to it.

NEC Terminology

There are two basic types of rules in the *NEC*: mandatory rules and advisory rules. Here is how to recognize the two types of rules and how they relate to all types of electrical systems.

Mandatory rules: All mandatory rules have the terms *shall* or *shall not* in them. The terms mean *must*. If a rule is mandatory, you *must* comply with it.

Permissive rules: All advisory rules have the terms *shall be permitted* or *shall not be required* in them. The terms in this case mean *recommended but not necessarily required.* If a rule is advisory, compliance is discretionary. If you want to comply with it, do so. But you don't have to if you don't want to.

Be alert to local amendments to the *NEC*. Local ordinances may amend the language of the *NEC*, changing it from *should* to *shall*. This means that you must do in that county or city what may only be recommended in some other area. The office that issues building permits will either sell you a copy of the code that's enforced in that county or tell you where the code is sold.

Learning the Layout of the NEC

Begin your study of the *NEC* with Articles 100 and 110. These two articles have the basic information that will make the rest of the *NEC* easier to understand. Article 100 defines terms you will need to understand the code. Article 110 gives the general requirements for electrical installations. Read these two articles over several times until you are thoroughly familiar with all the information they contain. It's time well spent.

Once you're familiar with Articles 100 and 110, you can move on to the rest of the code. There are several key sections you will use often in servicing electrical systems. Let's discuss each of these important sections.

Wiring Design and Protection

Chapter 2 of the *NEC* discusses wiring design and protection, the information electrical technicians need most often. It covers the use and identification of grounded conductors, branch circuits, feeders, calculations, services, overcurrent protection, grounding and surge protection. This is essential information for *any* type of electrical system, regardless of the type.

Chapter 2 is also a "how-to" chapter. It explains how to provide proper spacing for conductor supports and how to size the proper grounding conductor or electrode. If you run into a problem related to the design or installation of a conventional electrical system, you can probably find a solution for it in this chapter.

Wiring Methods and Materials

Chapter 3 has the rules on wiring methods and materials. The materials and procedures to use on a particular system depend on the type of building construction, the type of occupancy, the location of the wiring in the building, the type of atmosphere in the building or in the area surrounding the building, mechanical factors and the relative costs of different wiring methods.

The provisions of this article apply to all wiring installations except remote control switching (Article 725), low-energy power circuits (Article 725), signal systems (Article 725), communication systems and conductors (Article 800) when these items form an integral part of equipment such as motors and motor controllers.

There are three basic wiring methods used in most modern electrical systems. Nearly all wiring methods are a variation of one of these three basic methods:

- Sheathed cables of two or more conductors, such as NM cable and AC armored cable (Articles 320 through 340)
- Raceway wiring systems, such as rigid and EMT conduit (Articles 342 to 366)
- Busways (Article 368)

Article 310 in Chapter 3 gives a complete description of all types of electrical conductors. Electrical conductors come in a wide range of sizes and forms. Be sure to check the working drawings and specifications to see what sizes and types of conductors are required for a specific job. If conductor type and size are not specified, choose the most appropriate type and size meeting standard *NEC* requirements.

Articles 312 through 392 give rules for raceways, boxes, cabinets and raceway fittings. Outlet boxes vary in size and shape, depending on their use, the size of the raceway, the number of conductors entering the box, the type of building construction, and the atmospheric condition of the areas. Chapter 3 should answer most questions on the selection and use of these items.

The *NEC* does not describe in detail all types and sizes of outlet boxes But manufacturers of outlet boxes have excellent catalogs showing all of their products. Collect these catalogs. They're essential to your work.

Equipment for General Use

Chapter 4 of the *NEC* begins with the use and installation of flexible cords and cables, including the trade name, type, letter, wire size, number of conductors, conductor insulation, outer covering and use of each. The chapter also includes fixture wires, again giving the trade name, type, letter and other important details.

Article 404 covers the switches you will use to control electrical circuits.

Article 406 covers receptacles and convenience outlets used to connect portable equipment to electric circuits. Get the manufacturers' catalogs on these items. They will provide you with detailed descriptions of each of the wiring devices.

Article 408 covers switchboards and panelboards, including their location, installation methods, clearances, grounding and overcurrent protection.

Article 410 on lighting fixtures is especially important. It gives installation procedures for fixtures in specific locations. For example, it covers fixtures near combustible material and fixtures in closets. The *NEC* does not describe the number of fixtures that will be needed in a given area to provide a certain amount of illumination.

Article 430 covers electric motors, including mounting the motor and making electrical connections to it.

Articles 440 through 460 cover air conditioning and heating equipment, transformers and capacitors.

Article 480 gives most requirements related to battery-operated electrical systems. Storage batteries are seldom thought of as part of a conventional electrical system, but they often provide standby emergency lighting service. They may also supply power to security systems that are separate from the main AC electrical system.

Special Occupancies

Chapter 5 of the *NEC* covers *special occupancy* areas. These are areas where the sparks generated by electrical equipment may cause an explosion or fire. The hazard may be due to the atmosphere of the area or just the presence of a volatile material in the area. Commercial garages, aircraft hangers and service stations are typical special occupancy locations.

Articles 500 through 503 cover the different types of special occupancy atmospheres where an explosion is possible. The atmospheric groups were established to make it easy to test and approve equipment for various types of uses.

Section 501 covers the installation of explosionproof wiring. An explosionproof system is designed to prevent the ignition of a surrounding explosive atmosphere when arcing occurs within the electrical system.

There are three classes of special occupancy locations:

- Class I (Article 501): Areas containing flammable gases or vapors in the air. Class I areas include paint spray booths, dyeing plants where hazardous liquids are used, and gas generator rooms.
- Class II (Article 502): Areas where combustible dust is present, such as grain-handling and storage plants, dust and stock collector areas and sugar-pulverizing plants. These are areas where, under normal operating conditions, there may be enough combustible dust in the air to produce explosive or ignitable mixtures.
- Class III (Article 503): Areas that are hazardous because of the presence of easily ignitable fibers or flyings in the air, although not in large enough quantity to product ignitable mixtures. Class III locations include cotton mills, rayon mills and clothing manufacturing plants.

Articles 511 and 514 regulate garages and similar locations where volatile or flammable liquids are used. While these areas are not always considered critically hazardous locations, there may be enough danger to require special precautions in the electrical installation. In these areas, the *NEC* requires that volatile gases be confined to an area not more than 18 inches above the floor. So in most cases, conventional raceway systems are permitted above this level. If the area is judged critically hazardous, *explosionproof* wiring (including seal-offs) may be required.

Article 520 regulates theaters and similar occupancies where fire and panic can cause hazards to life and property. Drive-in theaters do not present the same hazards as enclosed auditoriums. But the projection rooms and adjacent areas must be properly ventilated and wired for the protection of operating personnel and others using the area.

Chapter 5 also covers residential storage garages, aircraft hangars, service stations, bulk storage plants, health care facilities, mobile homes and parks, and recreation vehicles and parks.

Special Equipment

Article 600 covers electric signs and outline lighting. Article 610 applies to cranes and hoists. Article 620 covers the majority of the electrical work involved in the installation and operation of elevators, dumbwaiters, escalators and moving walks. The manufacturer is responsible for most of this work. The electrician usually just furnishes a feeder terminating in a disconnect means in the bottom of the elevator shaft. The electrician may also be responsible for a lighting circuit to a junction box midway in the elevator shaft for connecting the elevator cage lighting cable and exhaust fans. Articles in Chapter 6 of the *NEC* give most of the requirements for these installations.

Article 630 regulates electric welding equipment. It is normally treated as a piece of industrial power equipment requiring a special power outlet. But there are special conditions that apply to the circuits supplying welding equipment. These are outlined in detail in Chapter 6 of the *NEC*.

Article 640 covers wiring for sound-recording and similar equipment. This type of equipment normally requires low-voltage wiring. Special outlet boxes or cabinets are usually provided with the equipment. But some items may be mounted in or on standard outlet boxes. Some sound-recording electrical systems require direct current, supplies from rectifying equipment, batteries or motor generators. Low-voltage alternating current comes from relatively small transformers connected on the primary side to a 120-volt circuit within the building.

Other items covered in Chapter 6 of the *NEC* include: X-ray equipment (Article 660), induction and dielectric heat-generating equipment (Article 665) and machine tools (Article 670).

If you ever have work that involves Chapter 6, study the chapter *before work begins*. That can save a lot of installation time. Here is another way to cut down on labor hours and prevent installation errors. Get a set of rough-in drawings of the equipment being installed. It is easy to install the wrong outlet box or to install the right box in the wrong place. Having a set of rough-in drawings can prevent those simple but costly errors.

Special Conditions

In most commercial buildings, the *NEC* and local ordinances require a means of lighting public rooms, halls, stairways and entrances. There must be enough light to allow the occupants to exit from the building if the general building lighting is interrupted. Exit doors must be clearly indicated by illuminated exit signs.

Chapter 7 of the *NEC* covers the installation of emergency lighting systems. These circuits should be arranged so that they can automatically transfer to an alternate source of current, usually storage batteries or gasoline-drive generators. As an alternative, you can connect them to the supply side of the main service so disconnecting the main service switch would not disconnect the emergency circuits. See Article 700.

How to Prepare for the Exam

This book is a guide to preparing for the journeyman or master electrician's exam. It isn't a substitute for studying the recommended references and it won't teach you the electrical trade. But it will give you a *complete knowledge of the type of questions* asked in the electrician's exam. It will also give you a "feel" for the examination and provide some of the confidence you need to pass.

Emphasis is on multiple-choice questions because that's the style that nearly all tests utilize. Questions are grouped into chapters. Each chapter covers a single subject. This will help you discover your strengths and weaknesses. Then when you take the two "final" sample exams in the back of this book, analyze the questions you miss. You'll probably notice you are weaker in some subjects than others. When these areas have been discovered, you will know that further study is necessary in these areas.

In answering questions on the *NEC*, remember this point: All exam questions are based on *minimum NEC requirements*. If the minimum wire size permitted under the *NEC* to carry 20 amperes is No. 12 AWG and you answer No. 10 AWG (minimum size for 30 amperes) just to play it safe, your answer is *incorrect*.

The preparatory questions in the front part of this book have the answer after each question. When reading a question, cover the answer with a card or ruler of an appropriate size. Read the question carefully. Mark your answer on a separate sheet of paper before moving the card or ruler that covers the correct answer. Then slide the card or ruler and check to see if your answer is correct. If it isn't, read the code responses to find out why it is wrong.

How to Study

Set aside a definite time to study, following a schedule that meets your needs. Studying a couple of hours two or three nights each week is better than studying all day on, say, Saturdays. The average mind can only concentrate for approximately four hours without taking a break. There's no point in studying if you don't retain much of the information. Study alone most of the time, but spend a few hours reviewing with another electrician buddy before exam day. You can help each other dig out the facts and concepts you will need to pass the exam.

Try to study in a quiet, well-lighted room that is respected as your study space by family members and friends. If it's hard to find a spot like that in your home, go to the local library where others are reading and studying.

Before you begin to study, spend a few minutes getting into the right frame of mind. That's important. You don't have to be a genius to pass the electrician's exam. But good motivation will nearly guarantee your success. No one can provide that motivation but *you*. Getting your license is a goal you set for yourself; it's your key to the future — a satisfying career in the electrical industry.

As you study the *NEC* and other references, highlight important point with a yellow marker. This makes it easier to find important passages when you're doing the final review — and when you're taking the exam.

Put paper tabs on the corners of each major section in all the references you will take into the exam room. On the portion of the tab that extends beyond the edge of the book, write the name of the section or the subject. That makes locating each section easier and quicker — an important consideration on an open book test. Speed in locating answers is important. In the sample exams at the end of this book, which are based on actual state and county examinations, you will have from two to four minutes to answer each question, so you don't have time to daydream or mess around. If you want to pass the exam, you must take it seriously.

Your study plan should allow enough time to review each reference at least three times. Read carefully the first time. The next review should take only about 10% of the time that the first reading took. Make a final review of all references and notes on the day before the exam. *This is the key to success in passing the exam: Review, review, review!* The more you review, the better your grasp of the information and the faster you will be able to find the answers.

The Examination

Questions on state and local examinations are usually compiled by members of the electrician's examination board. Board members usually include several electrical contractors, a registered electrical engineer, electrical inspectors, and perhaps a trade school instructor. Most electrician's exams will include questions on the *NEC*, general knowledge of electrical practice, theoretical questions, and local ordinance rules. All of these fields are covered in this preparation guide. Questions about the *NEC*, including rules and design calculations, comprise from 70% to 80% of the examination.

State examinations are usually given twice a year, or perhaps every three months. County and local exams may be taken almost any time with prior notice to the local inspectors. Most have several basic exams that are used in rotation. But the same examination will never be administered twice in a row.

The people compiling the exams maintain a bank of several hundred questions covering each test subject. Questions are selected at random, and chances are that some of the questions on any exam have already been used on an earlier examination. Many of the questions appearing on actual electrician's exams will closely resemble questions appearing in this book.

The format of the actual examination, the time allowed, and the reference material which the applicant may be allowed to take into the examination room vary with each locality. Typically, an applicant is allowed six to eight hours to complete the examination. Applicants are usually required to report to the examination room at 8 a.m. where the proctors take about 15 minutes to explain the rules of taking the exam. The applicants then work on the "morning" exam until noon. After an hour break for lunch, the "afternoon" exam begins at 1 p.m. and applicants are given until about 4 p.m. to complete this portion.

The Answer Sheet

Most answer sheets used today are designed for computer grading. Each question on the exam is numbered. Usually there will be four or five possible responses for each question. You will be required to mark the best answer on the answer sheet. The following is a sample of a multiple-choice question:

1) Richmond is the capitol city of what state?

(A) Florida (C) Virginia

(B) Maryland (D) California

You should mark answer *C* for question 1 on the answer sheet.

Answer sheets will vary slightly for each examining agency so be sure to follow any instructions on that sheet. Putting the right answers on the wrong section will almost certainly cause you to fail

The Night Before

Give your mind a rest! If you have not prepared correctly for the exam by this time, then you can't cram it all into your brain in one night. So take it easy. If the place of the examination is more than an hour's drive from your home, you might want to stay at a motel in the city where the examination is being held. Getting up at, say, 4 a.m. and driving a couple of hours in heavy traffic will not help you to pass the exam. On the other hand, a drive to the location the afternoon before the exam, a good dinner and a relaxing evening watching TV will increase your possibilities of passing. Just don't stay up too late.

There are, however, exceptions to this rule. Some people find it difficult to sleep comfortably the first night at a strange location. If this is your case, you would be better off getting a good night's sleep at home and driving to the location the next morning.

Just be sure to have all of your reference material with you and get a good night's sleep before the day of the exam. If you have prepared yourself correctly, you should pass with flying colors.

Examination Day

On the day of your examination, listen carefully to any oral instructions given and read the printed directions. Failing to follow instructions will probably disqualify you.

You will seldom find any trick questions, but many will require careful reading. Certain words (like *shall, should, always* and *never*) can make a big difference in your answer.

Sometimes several of the answers may seem possible, but only one will be correct. If you are not sure of the answer, use the process of elimination.

There are several ways to take an exam, but the following is the method I used to pass the Virginia State Electrical Contractor's Exam a few years ago. This method is merely a suggestion: if another way suits you best, by all means use it.

When the exam booklets were passed out and the applicants were given permission to open them, I spent the first few minutes going over the exam booklet, noting the number of questions. This allowed me to pace myself. I noted there were 100 questions on the morning exam — which gave me less than three minutes to spend on each one.

I then started with question No. 1. When I found one that I wasn't sure of, I merely skipped over this until I came to one that I definitely knew the answer. This way I had gone through the entire test booklet once and had answered about 50% of the questions in a little over one hour. I was quite sure that I had answered all of these questions correctly. However, 70% is usually the minimum passing grade and at this point, I had only 50% of the questions answered. But I still had about three hours to spend on the tougher questions.

I then started back at the beginning of the exam and went down the list of questions until I found one that was unanswered. This process continued until I had answered all questions to the best of my ability. I spent the remaining time reviewing my previous answers, making changes as necessary.

After lunch, the "afternoon" portion of the exam was handed out, and I used the same procedure as before. I found out a few days later that I had scored 94% on this examination.

What's New In This Edition?

All questions and answers in this book have been updated to comply with the new 2005 *NEC*. Additional questions and answers, along with new illustrations, have been provided to encompass new *NEC* installation requirements. Wherever a change has occurred from the 2002 *NEC*, you will see an icon denoting that a change has been made. This icon appears below.

Chapter 1

Electrical Systems — General Requirements

Owing to the potential fire and explosion hazards caused by the improper handling and installation of electrical wiring, certain rules in the selection of materials, quality of workmanship, and precautions for safety must be followed. To standardize and simplify these rules and provide a reliable guide for electrical construction, the National Electrical Code® (*NEC*) was developed. The *NEC*, originally prepared in 1897, is frequently revised to meet changing conditions, improved equipment and materials, and new fire hazards. It is the result of the best efforts of electrical engineers, manufacturers of electrical equipment, insurance underwriters, fire fighters, and other concerned experts throughout the country.

The *NEC* is now published by the National Fire Protection Association (NFPA), Batterymarch Park, Quincy, Massachusetts 02269. It contains specific rules and regulations intended to help in the practical safeguarding of persons and property from hazards arising from the use of electricity.

Although the *NEC* itself states, *This Code is not intended as a design specification nor an instruction manual for untrained persons,* it does provide a sound basis for the study of electrical design and installation procedures — under the proper guidance. The probable reason for the *NEC's* self-analysis is that the code also states, *This Code contains provisions considered necessary for safety. Compliance therewith and proper maintenance will result in an installation essentially free from hazard, but not necessarily efficient, convenient, or adequate for good service or future expansion of electrical use.*

The *NEC*, however, has become the bible of the electrical construction industry, and is usually the basis for most electrician's and electrical contractor's exams. Consequently, anyone involved in electrical work, in *any* capacity, should obtain an up-to-date copy, keep it handy at all times, and refer to it frequently.

To use the *NEC* properly, the definitions listed in Chapter 1, Article 100 of the *NEC* should be fully understood. General requirements for electrical installations are given in Article 110. Then, the remaining Chapters, Articles, and Sections should be studied.

1-1 "Concealed" as applied to electrical wiring means:

A) Rendered inaccessible by the structure or finish of the building

B) Capable of being reached quickly

C) Capable of being removed without damage

D) Admitting close approach

Answer: A

For example, cables or raceways installed within, say, a drywalled partition are not accessible without damaging the finished wall and are considered to be concealed. Wires in concealed raceways are considered concealed, even though they may become accessible by withdrawing them. NEC Article 100 — Definitions.

1-2 A feeder is:

A) A circuit conductor between the final overcurrent device protecting the circuit and the outlet

B) A branch circuit that supplies several outlets

C) All circuit conductors between the service equipment, the source of a separately derived system, or other power supply source and the final branch-circuit overcurrent device.

D) A device for generating electricity

Answer: C

NEC Article 100 — Definitions. See Figure 1-1.

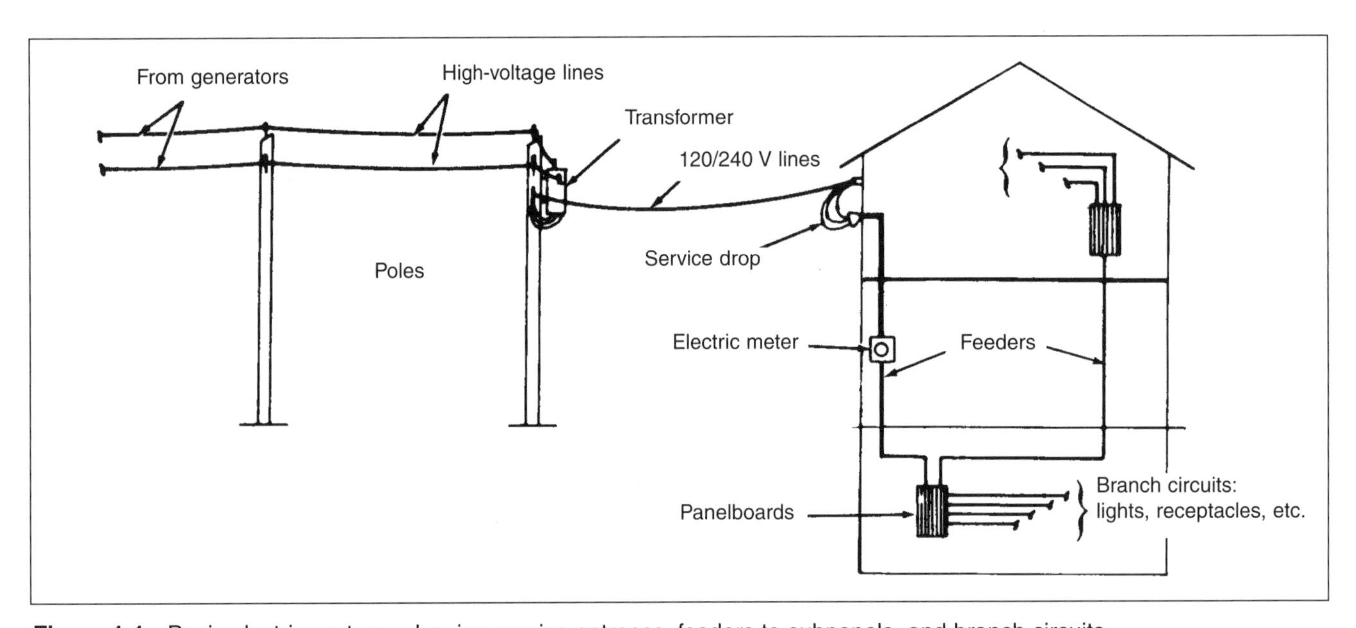

Figure 1-1: Basic electric systems showing service-entrance, feeders to subpanels, and branch circuits

1-3 A bonding jumper is:

A) A branch circuit that supplies only one utilization equipment

B) A reliable conductor to ensure the required electrical conductivity between metal parts required to be electrically connected

C) An adhesive used to insulate conductors

D) Capable of being operated without exposing operator to contact with live parts

Answer: B

NEC Article 100 — Definitions. See Figure 1-2.

1-4 In locations where electric equipment would be exposed to physical damage, the following must be provided:

A) Warning signs

B) Sufficient headroom

C) Working space

D) Enclosures or guards

Answer: D

NEC Section 110.27(B). See Figure 1-3.

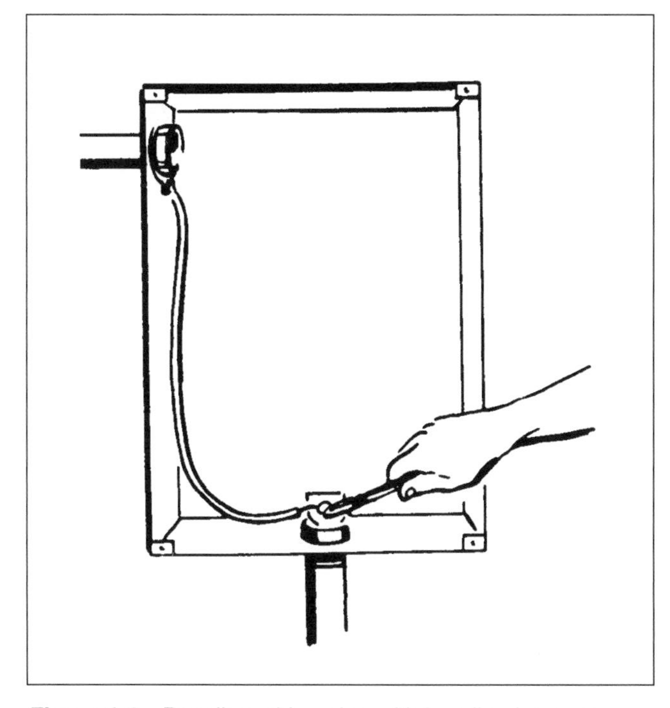

Figure 1-2: Panelboard housing with bonding jumpers

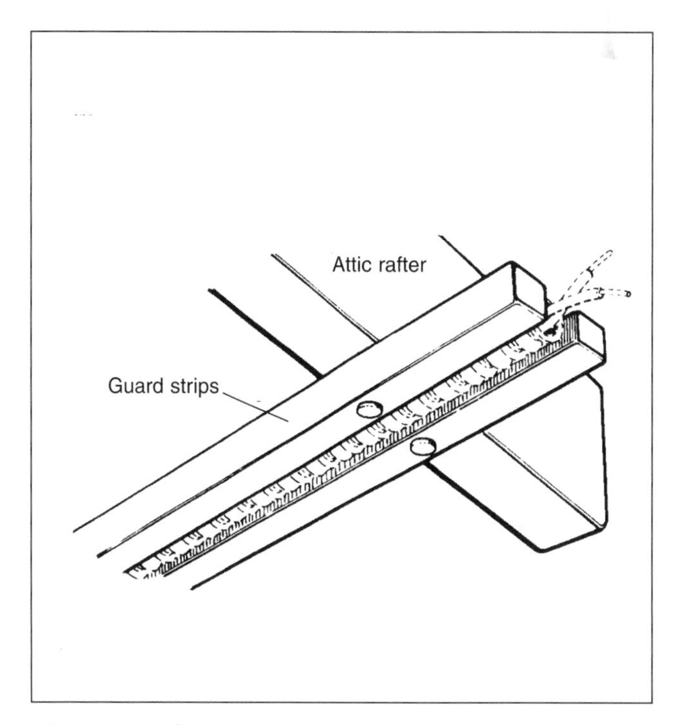

Figure 1-3: Guard strips protecting Type AC cable

1-5 To provide access to the working space about electric equipment, the following number of entrances of sufficient size must be provided:

A) 3

B) 2

C) 1

D) 4

Answer: C

For example, an electrical equipment room in an office building would require only one entrance door to the room. NEC Section 110.26(C).

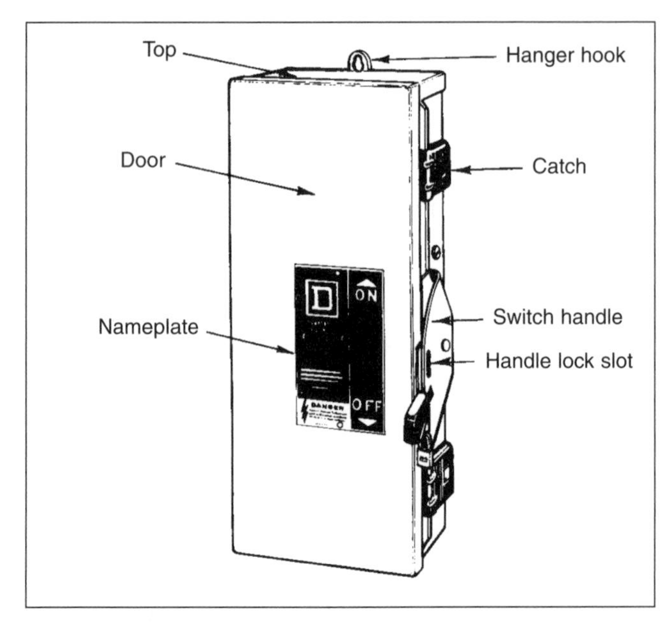

Figure 1-4: Externally-operated safety switch. The live interior parts are fully isolated, but the external handle enables the switch to be operated from the outside.

1-6 Externally operable means:

A) Capable of being operated from the outside of a building

B) An apparatus enclosed in a case

C) Capable of being operated without exposing the operator to contact with live parts

D) Surrounded by a case

Answer: C

NEC Article 100 — Definitions. See Figure 1-4.

1-7 What must be provided for in all working spaces above service equipment?

A) A water faucet to flush operator's eyes

B) A drinking fountain

C) Illumination

D) A wash basin

Answer: C

A light fixture must be installed for the working spaces about any switchboards, panelboards, etc. so adequate illumination (light) will be available for operation or repairs. NEC Section 110.26 (D).

1-8 Parts of electric equipment which in ordinary operation produce arcs, sparks, flames, or molten metal shall be enclosed or separated and isolated from:

A) All other electrical equipment

B) All combustible material

C) Electric lighting

D) All working spaces

Answer: B

NEC Section 110.18. (Special rules apply for motors and hazardous locations.)

1-9 Working space in rooms containing electrical equipment shall not be used for:

A) Storage

B) Maintenance and repair of equipment

C) Testing purposes

D) Inspection or servicing

Answer: A

The area around electrical equipment must be kept clear of foreign matter so that maintenance and repairs may be readily made. NEC Section 110.26 (B).

1-10 In all cases where there are live parts normally exposed on the front of switchboards or motor control centers, the working space in front of such equipment shall not be less than:

A) 1 foot

B) 3 feet

C) 4 feet unless adequate protection is provided

D) 18 inches

Answer: B

Three feet is judged by the NEC to be adequate space (600 volts or under) so that workers may keep a safe distance from live electrical parts. NEC Section 110.26(A) and NEC Table 110.26(A)(1).

1-11 Ampacity is defined as:

A) The electromotive force required to cause electrons to flow in conductors

B) The amount of power in a circuit

C) The current, in amperes, that a conductor can carry continuously under the conditions of use without exceeding its temperature rating

D) The voltage rating of any appliance

Answer: C

NEC Article 100 — Definitions.

1-12 "Approved" as used in the *NEC* means:

A) Acceptable to the authority having jurisdiction

B) Acceptable only when specified in local ordinances

C) Okay for use in hazardous locations

D) Usable only for inside work

Answer: A

In most cases, the local city, county, or state electrical inspector is the "authority" having jurisdiction. NEC Article 100 — Definitions.

1-13 A bare conductor is one with:

A) Several layers of thermoplastic insulation

B) Only one layer of insulation

C) A covering that is not recognized by the *NEC* as electrical insulation

D) No covering or insulation whatsoever

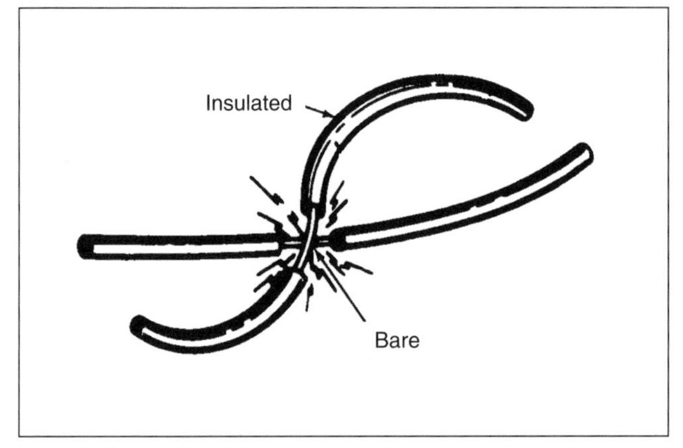

Figure 1-5: Bare conductor compared to one with insulation

Answer: D

A bare conductor, such as a service grounding wire, has no cover or insulation. NEC Article 100 — Definitions, Conductor, Bare. See Figure 1-5.

1-14 Which of the following anchors may not be used to secure electrical equipment to masonry walls?

A) Lead anchors approved for the weight of the equipment

B) Toggle bolts

C) Threaded studs "shot" into the masonry

D) Wooden plugs driven into holes in the masonry for holding wood screws

Answer: D

NEC Section 110.13(A). See Figure 1-6.

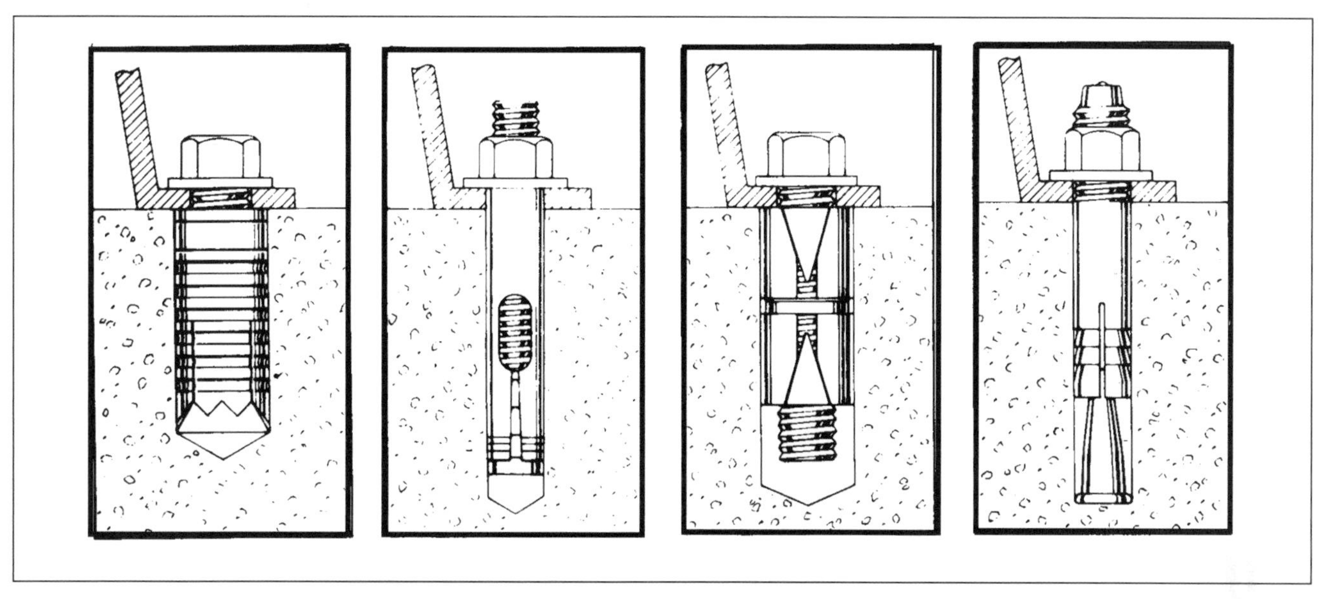

Figure 1-6: Several types of anchors suitable for installing electrical equipment

1-15 A device designed to open and close a circuit by nonautomatic means and to open the circuit automatically on a predetermined overcurrent without damage to itself when properly applied within its rating is called a:

A) Nonfusible disconnect switch

B) Time-delay fuse

C) Circuit breaker

D) Motor running overcurrent protector

Answer: C

NEC Article 100 — Definitions. See Figure 1-7.

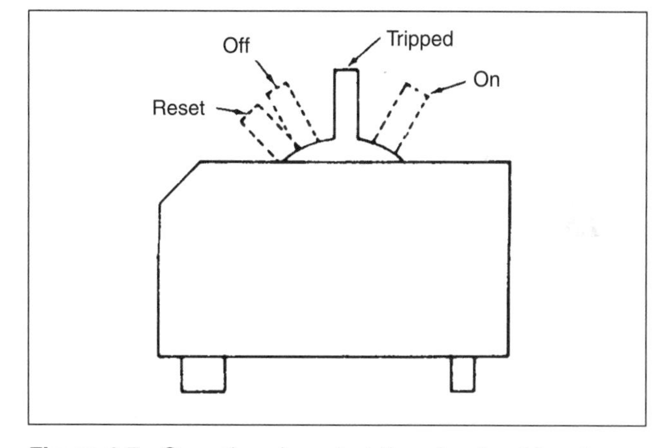

Figure 1-7: Operating characteristics of a circuit breaker

1-16 A conductor encased within material of composition or thickness that is recognized by the *NEC* as electrical insulation is known as a:

A) Bare conductor

B) Covered conductor

C) Concealed conductor

D) Insulated conductor

Answer: D

NEC Article 100 — Definitions, Conductor, Insulated.

1-17 A separate portion of a conduit or tubing system that provides access through a removable cover(s) to the interior of the system at a junction of two or more sections of the system or at a terminal point of the system is defined as a:

A) Conduit body

B) Conduit junction

C) Conduit intersection

D) Conduit T-connector

Answer: A

NEC Article 100 — Definitions. See Figure 1-8.

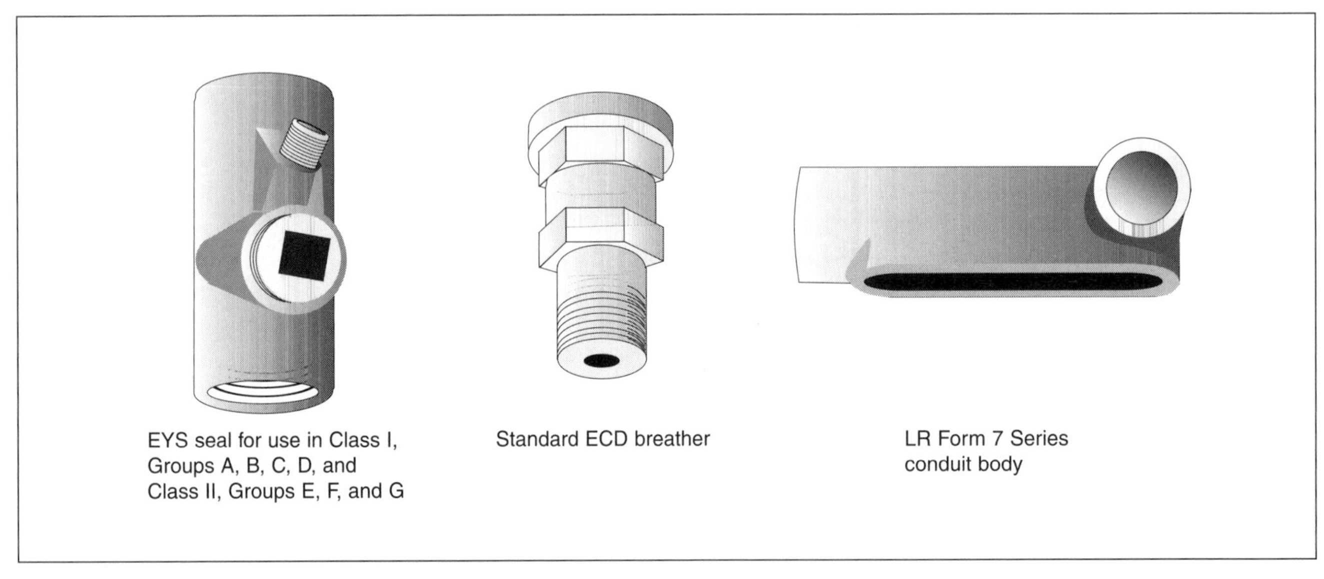

Figure 1-8: Several types of conduit bodies

1-18 Which of the following describes cleaning and lubricating compounds that can cause severe damage to many plastic insulating materials?

A) Deteriorating agents

B) Cooling agents

C) Splicing agents

D) Heating agents

Answer: A

NEC Section 110.11 deals with deteriorating agents, stating that some cleaning and lubricating compounds can cause severe deterioration of many plastic materials used for insulating and structural applications in equipment.

1-19 Which circuits must not be connected to any system containing trolley wires with a ground return?

A) Ground wires

B) Grounding conductors

C) Circuits for lighting and power

D) Ungrounded trolley wires

Answer: C

No other "live" or ungrounded conductors may be connected except those specifically designed for trolley operation. NEC Section 110.19.

1-20 The minimum headroom of working spaces about service equipment, switchboards, panelboards, or motor control centers shall be at least:

A) 6½ feet

B) 8 feet

C) 5.75 feet

D) 10 feet

Answer: A

The NEC judges 6½ feet working space to be adequate above service equipment to provide room for an electrician to service the equipment. This measurement is taken from the floor to ceiling of, say, an electrical equipment room; not from the top of the panelboard. Where the electrical equipment exceeds the minimum dimensions, the minimum headroom must not be less than the height of the equipment. NEC Section 110.26(E).

1-21 Indoor electrical installations over 600 volts that are open to unqualified persons shall be made with:

A) Open switchgear with readily accessible live parts

B) Metal-enclosed equipment

C) Provisions to enclose the equipment within a barrier less than 8 feet high

D) The approval of both the I.B.E.W. and the I.E.S.

Answer: B

Equipment shall be metal-enclosed, or in an area to which access is controlled by a lock. NEC Section 110.31(B)(1).

1-22 The entrance provided to give entrance and access to the working space about electric equipment rated over 600 volts must not be less than:

A) 6 feet × 6 feet

B) 24 feet × 6 feet

C) 2 feet × 6 feet

D) 2 feet × 6½ feet

Answer: D

Additionally, door(s) shall open in the direction of egress and be equipped with panic bars, pressure plates, or other devices that are normally latched but open under simple pressure. NEC Section 110.33(A).

1-23 Where switches, or other equipment operating at 600 volts, nominal, or less, are installed in a room or enclosure where there are exposed energized parts or wiring operating at over 600 volts, the high-voltage equipment shall be effectively separated from the space occupied by the low-voltage equipment by a suitable:

A) Warning sign

B) Partition, fence or screen

C) Voltage-reducing transformer

D) Firewall

Answer: B

A partition to prevent contact between the two systems must be installed. NEC Section 110.34(B).

1-24 An enclosed channel designed expressly for holding wires, cables, or busbars is called:

A) A hose

B) A raceway

C) A receptacle

D) A panelboard

Answer: B

A raceway may be conduit or piping, auxiliary wire troughs (gutters), busducts, wire trays, etc. NEC Article 100 — Definitions.

1-25 A contact device installed at an outlet for the connection of an attachment plug is called:

A) A terminator

B) A junction box

C) An overcurrent protection device

D) A receptacle

Answer: D

NEC Article 100 — Definitions.

1-26 A compartment or chamber to which one or more air ducts are connected and which forms part of the air distribution system is a:

A) Plenum

B) Duct

C) Fan-coil unit

D) Air valve

Answer: A

NEC Article 100 — Definitions. See Figure 1-9.

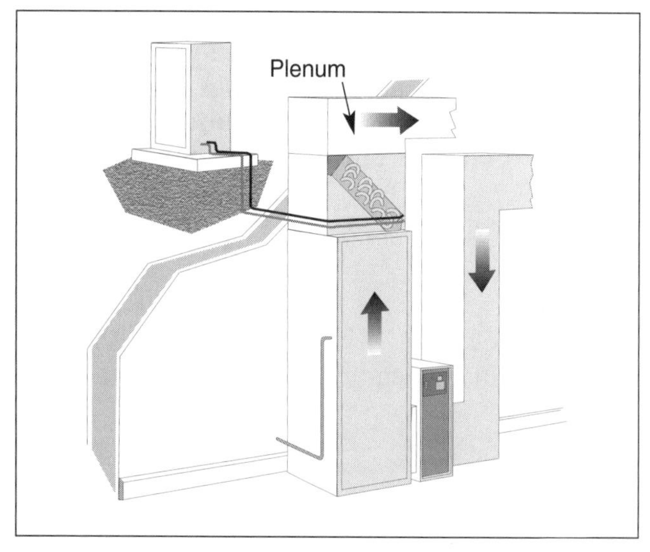

Figure 1-9: An air-distribution system showing a plenum and related air ducts

1-27 The *NEC* definition of "Qualified Person" is one who:

A) Has been elected by the Inspection office

B) Has skills and knowledge related to the construction and operation of electrical equipment and installations and has received safety training on the hazards involved.

C) Has served two years apprenticeship training with a labor organization

D) Has a college degree in electrical engineering or electrical technology

Answer: B

A licensed electrician and a professional electrical engineer are two such "qualified" persons. NEC Article 100 — Definitions.

1-28 When a component used on an electrical wiring system has been constructed or treated to prevent rain from interfering with the successful operation of the apparatus under specified test conditions, it is known as:

A) Raintight

B) Readily accessible

C) Rainproof

D) Rated-load component

Answer: C

A rainproof device may be used outside in rainy conditions. A rainproof device or apparatus may allow rain water to enter the enclosure, but such rain will not interfere with its operation. Raintight devices or enclosures are designed to keep the rain out of the enclosure. NEC Article 100 — Definitions.

1-29 When an enclosure has been constructed so that exposure to a beating rain will not result in the entrance of water under specified test conditions, the enclosure is known as:

A) Rainproof

B) Drip-proof

C) Raintight

D) Multioutlet assembly

Answer: C

This type of enclosure is designed to keep rain water out of the device. NEC Article 100 — Definitions. See Figure 1-10.

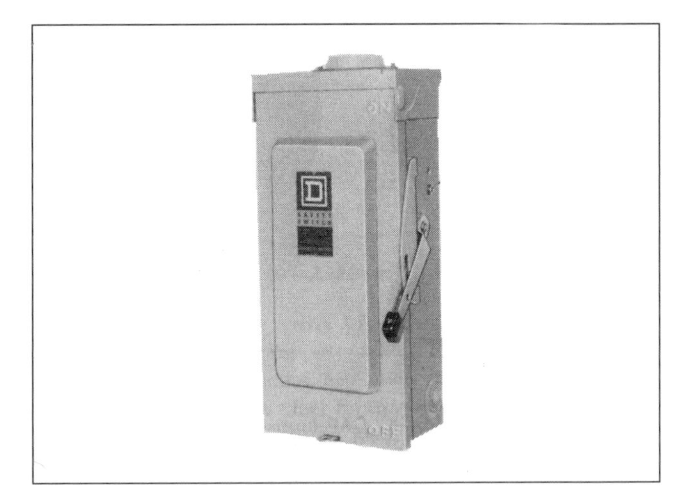

Figure 1-10: A typical raintight enclosure

1-30 A point on the wiring system at which current is taken to supply utilization equipment is known as:

A) An outlet

B) A wall switch

C) A load center

D) A motor-control center

Answer: A

A duplex receptacle, range receptacles, dryer receptacles, lighting fixtures, etc. are all outlets. NEC Article 100 — Definitions.

1-31 The agency that publishes the *NEC* is abbreviated:

A) HVAC

B) EEEC

C) NFPA

D) NRA

Answer: C

The National Fire Protection Association.

1-32 A raceway system is an enclosed channel designed expressly for holding:

A) Wires, cables or busbars

B) Automotive equipment

C) Spark-plug wires

D) Equipment for high-speed autos

Answer: A

A conduit, such as rigid steel conduit, and unlike plumbing pipe, is designed especially for containing electrical conductors. NEC Article 100 — Definitions.

1-33 A large single panel, or assembly of panels, on which are mounted switches, overcurrent and other protective devices will fall under the definition of:

A) General-use switch

B) Thermal protector

C) Switchboard

D) Cutout

Answer: C

Switchboards are generally accessible from the rear as well as from the front and are not intended to be installed in cabinets. NEC Article 100 — Definitions. See Figure 1-11.

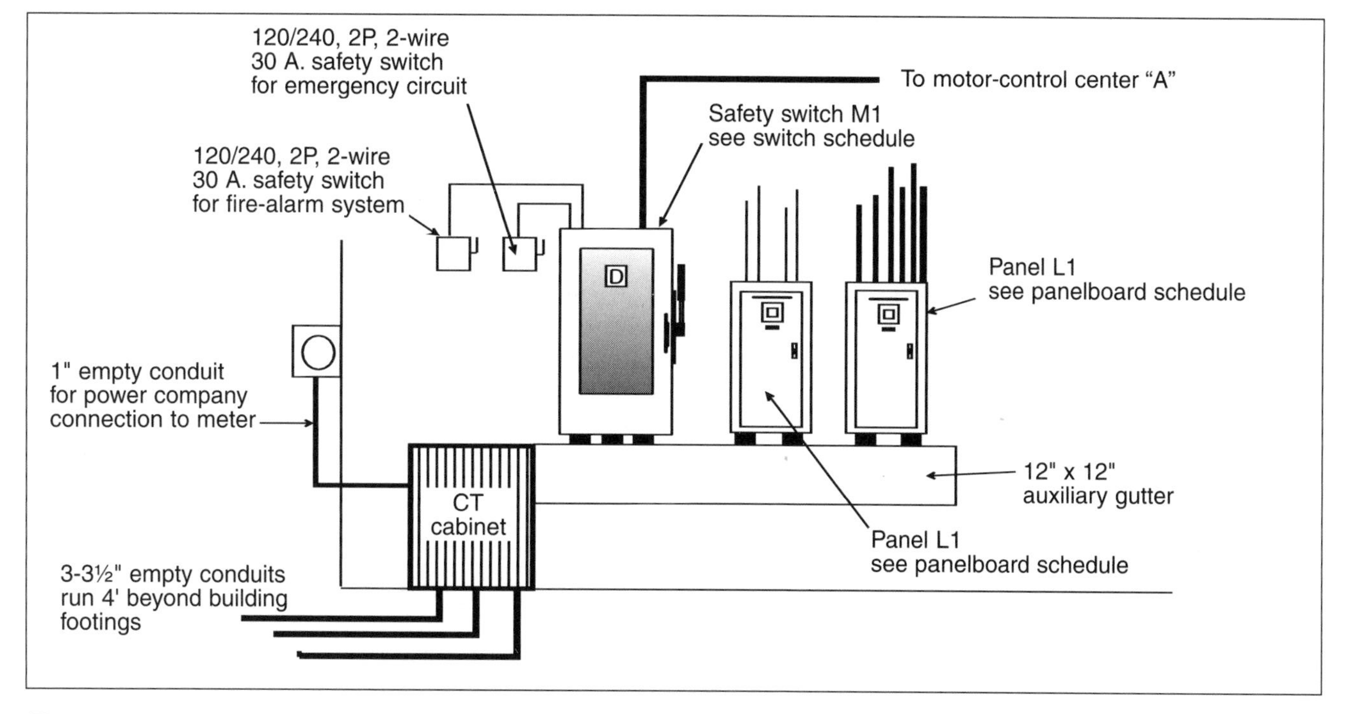

Figure 1-11: An assembly of safety switches, auxiliary gutters, and conduit nipples used for service equipment exactly as they appear on electrical working drawings

1-34 Any electric circuit that controls any other circuit through a relay is called a:

A) Remote-control circuit

B) Power circuit

C) Overload relay

D) Motor control circuit

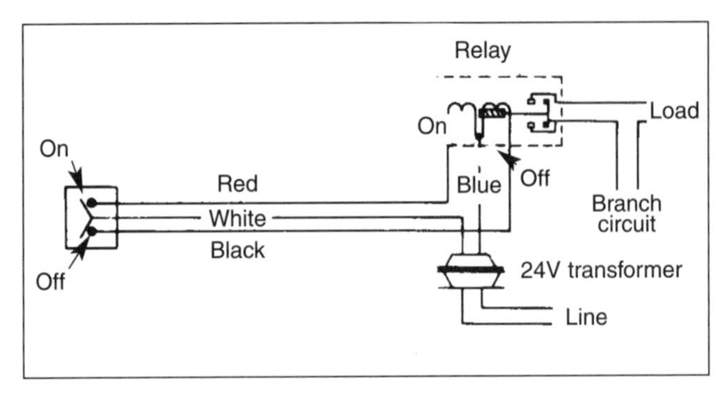

Figure 1-12: Typical remote-control circuit

Answer: A

NEC Article 100 — Definitions. See Figure 1-12.

1-35 Any electric circuit that energizes signaling equipment is known as a:

A) Low-voltage branch circuit

B) Multiwire circuit

C) Signaling circuit

D) Feeder circuit

Answer: C

For example, a circuit consisting of a low-voltage transformer, pushbuttons, door chime, and conductors is a signaling circuit; so is a burglar alarm system. NEC Article 100 — Definitions.

1-36 A device which, by insertion in a receptacle, establishes connection between the conductor of the attached flexible cord and the conductors connected permanently to the receptacle is called one of the following:

A) Female plug

B) Circuit breaker

C) Controller

D) Attachment plug

Answer: D

Also called a plug cap or plug. NEC Article 100 — Definitions. See Figure 1-13.

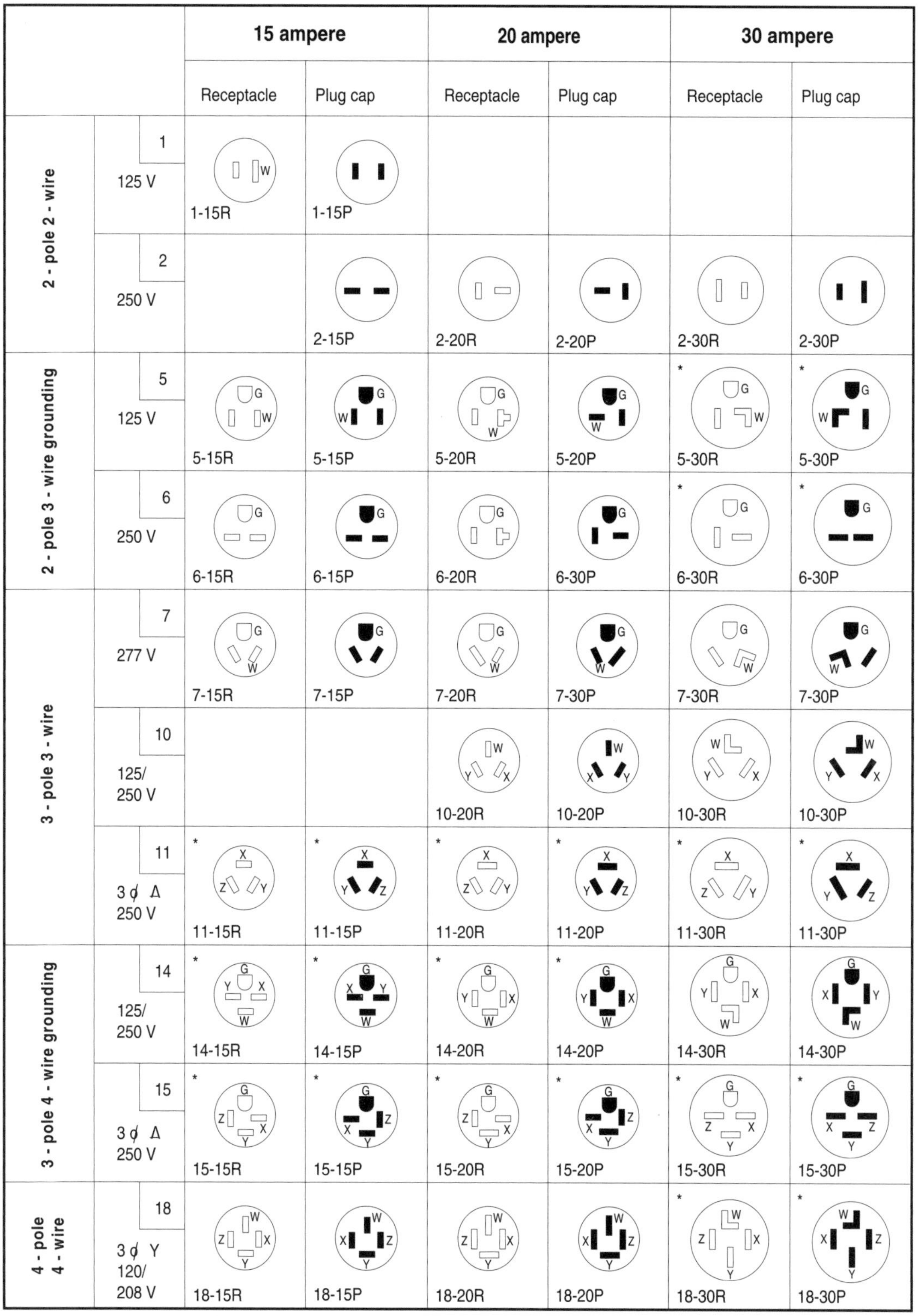

		15 ampere		20 ampere		30 ampere	
		Receptacle	Plug cap	Receptacle	Plug cap	Receptacle	Plug cap
2 - pole 2 - wire	1 125 V	1-15R	1-15P				
	2 250 V		2-15P	2-20R	2-20P	2-30R	2-30P
2 - pole 3 - wire grounding	5 125 V	5-15R	5-15P	5-20R	5-20P	* 5-30R	* 5-30P
	6 250 V	6-15R	6-15P	6-20R	6-30P	* 6-30R	* 6-30P
3 - pole 3 - wire	7 277 V	7-15R	7-15P	7-20R	7-30P	7-30R	7-30P
	10 125/ 250 V			10-20R	10-20P	10-30R	10-30P
	11 3ϕ Δ 250 V	* 11-15R	* 11-15P	* 11-20R	* 11-20P	* 11-30R	* 11-30P
3 - pole 4 - wire grounding	14 125/ 250 V	* 14-15R	* 14-15P	* 14-20R	* 14-20P	14-30R	14-30P
	15 3ϕ Δ 250 V	* 15-15R	* 15-15P	* 15-20R	* 15-20P	* 15-30R	* 15-30P
4 - pole 4 - wire	18 3ϕ Y 120/ 208 V	18-15R	18-15P	18-20R	18-20P	* 18-30R	* 18-30P

Figure 1-13: Several types of attachment-plug configurations

1-37 A device that establishes a connection between two or more conductors by means of mechanical pressure and without the use of solder is called:

A) An explosionproof connector with seal-offs

B) A pressure connector

C) A wire nut

D) A shrink connector

Answer: B

NEC Article 100 — Definitions, Connector, Pressure. See Figure 1-14.

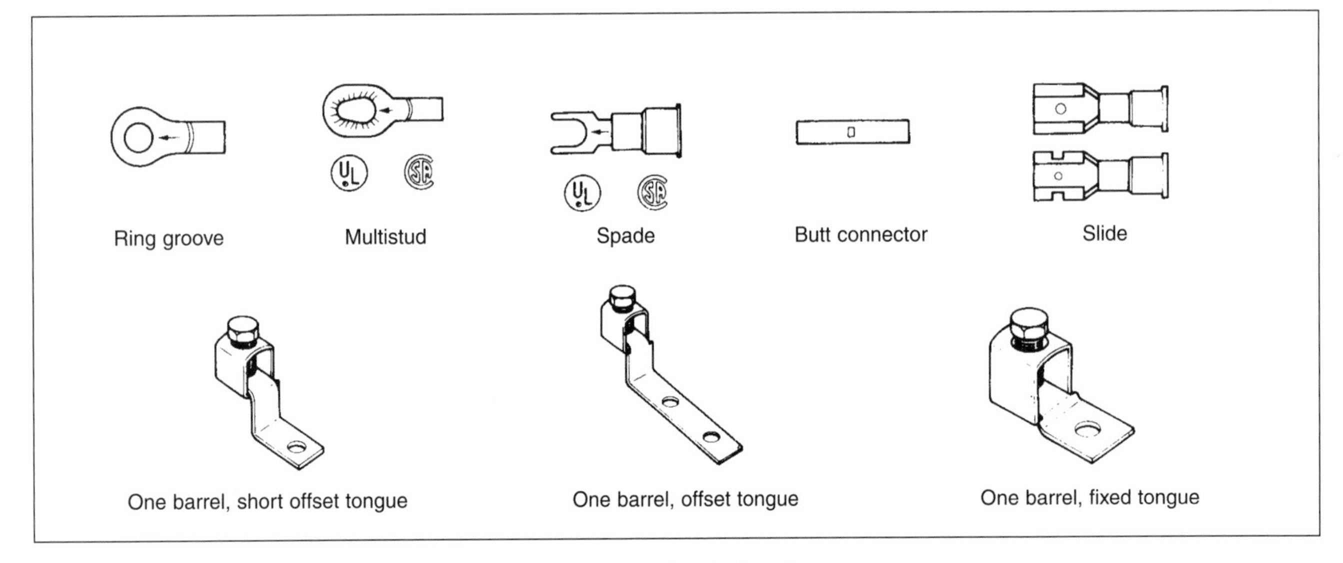

Figure 1-14: Several types of pressure connectors used in electrical work

1-38 A continuous load is a load where the maximum current is expected to continue for a certain length of time. This time is:

A) One hour or more

B) Two hours or more

C) Three hours or more

D) Four hours or more

Answer: C

For example, electric baseboard heaters in the coldest weather will more than likely operate for longer than three hours. Therefore, circuits feeding these units must be rated as "continuous." NEC Article 100 — Definitions.

1-39 A branch circuit that supplies a number of outlets for lighting and appliances is what type of circuit?

A) An appliance branch circuit

B) An individual branch circuit

C) A general purpose branch circuit

D) A multiwire branch circuit

Answer: C

A branch circuit that supplies two or more receptacles or outlets for lighting and appliances. Circuits feeding any duplex receptacles other than small appliance and laundry equipment are also general purpose branch circuits. NEC Article 100 — Definitions, Branch Circuit, General Purpose.

1-40 A device used to govern, in some predetermined manner, the electric power delivered to an electric apparatus is called a:

A) Controller

B) Heater

C) Governor

D) Motor starter

Answer: A

A wall switch controlling a lighting fixture is one. A motor starter or controller is another example. A rheostat or dimmer used to vary the light intensity is also a controller. NEC Article 100 — Definitions.

1-41 Which of the following qualifying terms indicate that a circuit breaker can be set to trip at various values of current, time, or both, within a predetermined range?

A) Accessible

B) Adjustable

C) Setting

D) Concealed

Answer: B

The term "adjustable" (as applied to circuit breakers) means that the circuit breaker can be set to trip at various values of current, time, or both, within a predetermined range. NEC Article 100 — Definitions, Circuit Breaker, Adjustable.

1-42 The ratio of the maximum demand of a system, or part of a system, to the total connected load of the system or the part of the system under consideration is known as:

A) Percentage

B) Duty cycle

C) Rated-load current

D) Demand factor

Answer: D

The NEC recognizes that every electrical outlet or piece of electric equipment will not all be operating simultaneously. Therefore, the NEC allows a demand factor for certain installations. NEC Article 100 — Definitions.

1-43 Electric parts that are not suitably guarded, isolated, or insulated and are capable of being inadvertently touched or approached nearer than a safe distance by a person are known as:

A) Exposed

B) Externally operable

C) Accessible

D) Dead front

Answer: A

A switchboard with knife switches, for example, has exposed live electrical parts. NEC Article 100 — Definitions.

1-44 A device intended for the protection of personnel that functions to de-energize a circuit or portion thereof within an established period of time when a current to ground exceeds the values established for a class A device.

A) Grounding electrode device

B) Ground-fault circuit-interrupter

C) Guarded protector

D) Thermal cutout

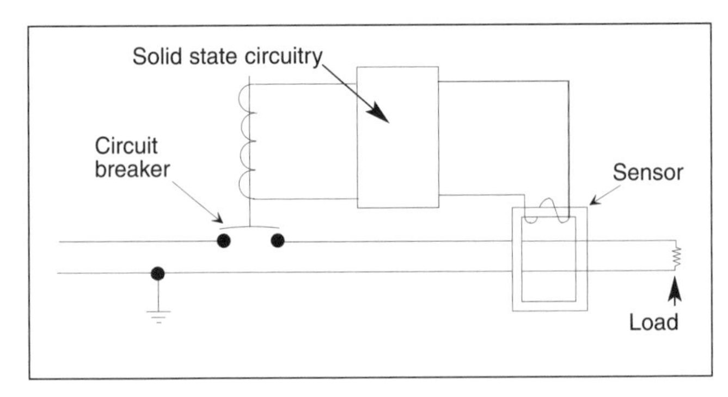

Figure 1-15: Ground-fault circuit-interrupter circuit

Answer: B

Class A GFCIs trip when the current to ground has a value in the range of 4-6 mA. Ground-fault circuit-interrupters are required on all residential receptacles installed outdoors, in bathrooms, and in garages. NEC Article 100 — Definitions. See Figure 1-15.

1-45 The *NEC* uses the term "isolated" to mean:

A) Not readily accessible to persons unless special means for access are used

B) Grouped together

C) Identifiable by means of color coding or nameplate

D) Nearby

Answer: A

For example, a safety switch with a means of locking the access door to live interior parts would be considered to be not readily accessible. NEC Article 100 — Definitions.

1-46 Accessories such as locknuts, bushings, etc. are known as:

A) Connectors

B) Conduit bodies

C) Fittings

D) Ground clips

Answer: C

A fitting can be a locknut, bushing, or other part of a wiring system that is intended to perform a mechanical rather than an electrical function. NEC Article 100 — Definitions.

1-47 An arrangement of incandescent lamps or electric discharge lighting to call attention to certain features such as the shape of a building is called:

A) Festoon lighting used for outdoor parties and to highlight other outdoor functions

B) Outline lighting

C) High-intensity discharge lighting such as normally used at intersections of highways

D) Decorative lighting inside a building

Answer: B

Lighting fixtures of many types, but especially neon, are used to highlight signs, buildings, and the like. Las Vegas gambling casinos are a good example of how outline lighting is used to highlight and outline buildings. NEC Article 100 — Definitions.

1-48 Operation of equipment in excess of normal, full-load rating is known as:

A) Hot load

B) Under current

C) Overload

D) Periodic duty

Answer: C

For example, a conductor operating in excess of its rated ampacity that, if it persists for a sufficient length of time, would cause damage or dangerous overheating. A fault, such as a short circuit or ground fault, is not an overload. NEC Article 100 — Definitions.

1-49 The value of current, time, or both at which an adjustable circuit breaker is set to trip is known as:

A) Inverse time

B) Ampacity

C) Automatic

D) Setting

Answer: D

The setting of a circuit breaker is the value of current, time, or both at which an adjustable circuit breaker is set to trip. NEC Article 100 — Definitions, Circuit Breaker, Setting.

1-50 Which of the following must not be allowed to come in contact with interior parts of electrical equipment?

A) Busbars

B) Wiring terminals

C) Abrasives

D) Insulators

Answer: C

NEC Section 110.12(C) states that internal parts of electrical equipment must not be damaged or contaminated by foreign materials such as paint, plaster, cleaners, abrasives or corrosive residues.

1-51 How tall must a wall, screen, or fence be that encloses an outdoor electrical installation over 600 volts to deter access by unqualified persons?

A) 5 feet

B) 6 feet

C) 7 feet

D) 8 feet

Answer: C

NEC Section 110.31 requires a wall, screen, or fence to be not less than 7 feet in height when it is used to enclose an outdoor electrical installation with voltages over 600 volts.

1-52 Entrances to all rooms or other enclosures containing exposed live parts operating at over 600 volts, nominal, shall be:

A) Elevated 30 feet above ground

B) Painted yellow with black and blue stripes

C) Kept unlocked for immediate servicing

D) Kept locked

Answer: D

NEC Section 110.34(C) requires that entrances to all buildings, rooms, or enclosures containing exposed live parts or exposed conductors operating at over 600 volts, nominal, be kept locked unless such entrances are under the observation of a qualified person at all times.

1-53 The *NEC* definition for a "bathroom" is an area containing one or more of the following: toilet, tub, or shower, and at least one:

A) Vent-through-roof

B) Hand dryer

C) Electric heater

D) Basin

Answer: D

NEC Article 100 — Definitions requires at least one basin and one or more toilets, tubs, or showers for an area to be classified as a bathroom.

1-54 Which of the following best describes festoon lighting?

A) A string of outdoor lights that is suspended between two points

B) A 600 volt lighting system used in power plants

C) Low voltage lighting used on billboards

D) Underwater lighting in pools and ponds

Answer: A

Defined in NEC Article 100 — Definitions.

1-55 What is the term used to describe an enclosing case that will not allow dust to enter under specified test conditions?

A) Dustproof

B) Dust prevention

C) Dusttight

D) Explosionproof

Answer: C

NEC Article 100 — Definitions.

1-56 What is an assembly of one or more enclosed sections having a common power supply (bus) and principally containing motor control units called?

A) Main distribution panel

B) Motor control center

C) Motor load center

D) Selector-switch control center

Answer: B

NEC Article 100 — Definitions states that the described apparatus is referred to as a "motor control center." See Figure 1-16.

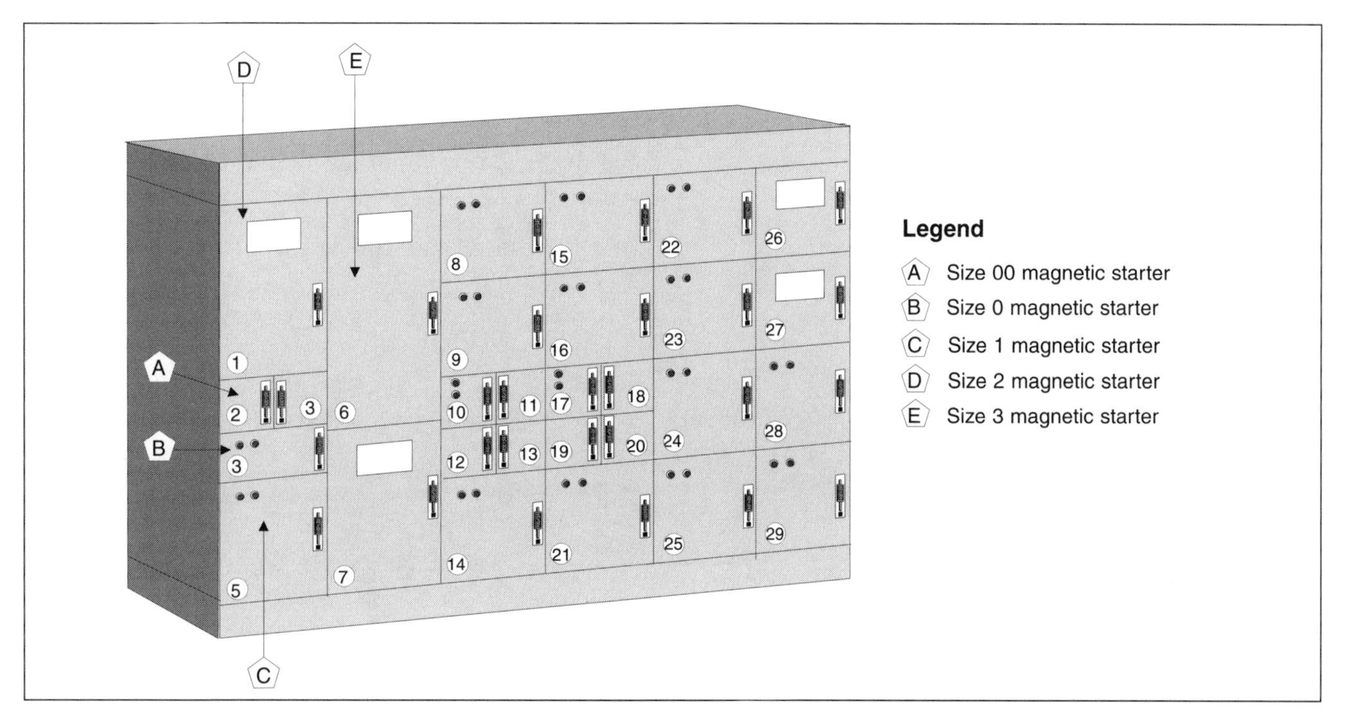

Figure 1-16: Typical motor control center

1-57 What is the name of an electrical load where the wave shape of the steady-state current does not follow the wave shape of the applied voltage?

A) Linear

B) Parallel load

C) Nonlinear load

D) Series load

Answer: C

NEC Article 100 — Definitions.

1-58 How many access entrances are required to working space about electrical equipment rated 1200 amperes or more and over 6 feet wide?

A) One

B) Two, one at each end of the area

C) Three, one on each of three sides

D) Four, one on each of four sides

Answer: B

NEC Section 110.26(C) requires one entrance not less than 24 inches wide and 6½ feet high at each end of the work space. Additional requirements apply to the installation and hardware requirements of such doors.

1-59 Where the electrical equipment exceeds 6½ feet in height, what is the required minimum headroom?

A) One foot higher than the top of the highest piece of equipment

B) Not less than the height of the equipment

C) 6 inches more than the height of the equipment

D) 4 feet more than the height of the equipment

Answer: B

NEC Section 110.26(E) requires a minimum headroom working space of 6½ feet; the minimum headroom shall not be less than the height of the equipment if the equipment exceeds 6½ feet in height.

1-60 Where must branch circuits and feeders be identified?

A) At the disconnecting means

B) At the point where they terminate

C) At both the point where they originate and the point where they terminate

D) Midway between the outlet and overcurrent protective device

Answer: A

NEC Section 110.22 requires circuit identification at the disconnecting means. This is usually at the main distribution panel for feeders and at subpanels for branch circuits. If overcurrent protective devices are located properly, this is the location where the circuits should be identified.

1-61 What is the minimum depth of clear working space at electrical equipment rated above 75 kV and classified as Condition 1?

A) 3 feet

B) 5 feet

C) 6 feet

D) 8 feet

Answer: D

NEC Table 110.34(A) requires 8 feet minimum depth of clear working space under the conditions described in this question.

Chapter 2

Electrical Calculations

Calculations are required for any type of electrical work — from design to installations. Consequently, anyone involved in any phase of electrical work — in any capacity — will frequently be called upon to make certain mathematical calculations. Therefore, most electrician's examinations will have some electrical problems involving the knowledge of equations and mathematical calculations. The examples to follow were taken from actual county and state examinations. They are typical of those that will be encountered on all Master Electrician's Exams and also for State Electrical Contractor's Exams.

The basic math operations are multiplication, division, addition, and subtraction. My high school algebra teacher would frequently remind her students of this by using the phase, "My Dear Aunt Sally." Furthermore, she would frequently remind us that regardless of how complicated the equation, you could not do more than use these four operations. Electricians should be able to use these operations in solving whole number problems, decimal number problems, and problems dealing with fractions. With these four basic math operations, along with squares, square roots, percents, and solving of equations, the electrician will have all the basic tools necessary for most electrical calculations.

It is beyond the scope of this book to review mathematical functions, other than the examples given. Therefore, if you feel that your knowledge of mathematics is lacking, it is highly recommended that you take some means to improve your knowledge before taking the examination. In general, however, the electrician can solve most electrical equations if he or she isolates the unknown quantity on one side of an equation by using the undoing method to move numbers from the side with the unknown quantity. For example, see Figure 2-1.

$R_1 = 20\Omega$, $R_2 = 30\Omega$, total resistance = 10Ω

To find R_3, place the values in the parallel resistor equation.

$$\frac{1}{R_t} = \frac{1}{R_1} + \frac{1}{R_2} + \frac{1}{R_3}$$

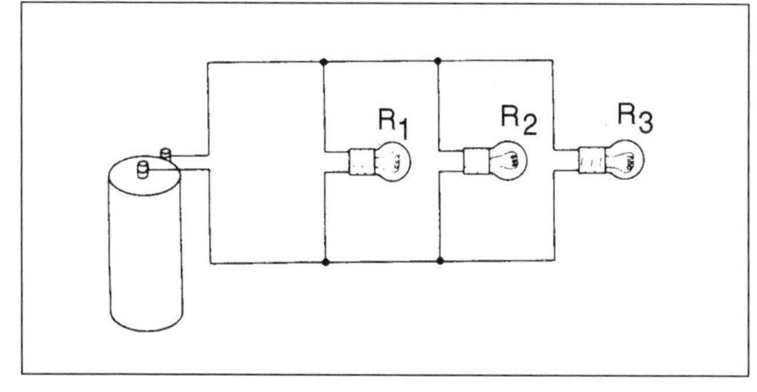

Figure 2-1: Typical electric circuit with three resistors connected in parallel

Substituting the known values in the equation, you get:

$$\frac{1}{10}=\frac{1}{20}+\frac{1}{30}+\frac{1}{R_3}$$

Then subtracting from $\frac{1}{20}+\frac{1}{30}$ both sides of the equation, you get:

$$\frac{1}{10}-\frac{1}{20}-\frac{1}{30}=\frac{1}{R_3}$$

Find common denominators:

$$\frac{6}{60}-\frac{3}{60}-\frac{2}{60}=\frac{1}{R_3}$$

$$\frac{1}{60}=\frac{1}{R_3}$$

Since this equation is a proportion, multiply diagonally as follows:

$$1 \text{ x } R_3 = (60)(1)$$

$$R_3 = 60 \text{ ohms}$$

As long as the values of all variables are known, and you have the equation that relates the variables, you should be able to use equation-solving techniques such as the undoing process to solve such equations.

Once you have gone over the questions in this chapter, you should have a good knowledge of the types of questions that will appear on the typical electrician's examination. You should also be aware of the areas where you are the weakest so that steps may be taken to improve your knowledge.

2-1 The circuit in Figure 2-2 shows the resistors R_2 and R_3 connected in parallel, but resistor R_1 is in series with both the battery and the parallel combination of R_2 and R_3. What is the total resistance of R_2 and R_3?

A) 40 ohms

B) 10 ohms

C) 20 ohms

D) 30 ohms

Answer: B

$$R_p = \frac{R}{n} = \frac{20}{2} = 10\ ohms$$

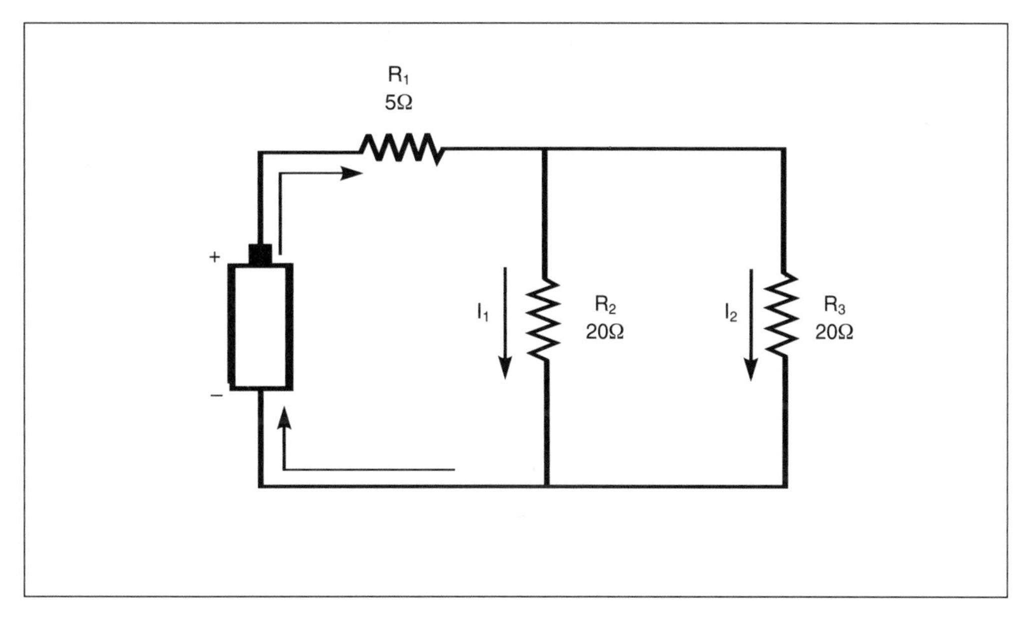

Figure 2-2: Electrical circuit with resistors connected in both series and parallel

2-2 Now we know that the total combined resistance of R_2 and R_3 is 10 ohms. What is the total resistance of the entire circuit in Figure 2-2?

A) 15 ohms

B) 20 ohms

C) 45 ohms

D) 5 ohms

Answer: A

$R_t = R_1 + R_p = 15\ ohms$

2-3 Assuming the battery in Figure 2-2 is 30 volts, what is the total current in the circuit?

A) 1 ampere
B) 2 amperes
C) 3 amperes
D) 4 amperes

Answer: B

$$I_t = E / R_t = \frac{30}{15} = 2 \text{ amperes}$$

2-4 According to Kirchhoff's voltage law, which of the following Ohm's law equations may be used to find the voltage drop across a resistor?

A) Voltage drop = IR
B) Voltage drop = I/R
C) Voltage drop = E/R
D) Voltage drop = W/I

Answer: A

$E \text{ (voltage)} = I \text{ (current)} \times R \text{ (resistance)}$

2-5 A 120-volt circuit has an electric heater connected with a current rating of 7.5 amperes. What is the resistance in ohms of the connected pure resistance load?

A) 8 ohms
B) 10 ohms
C) 16 ohms
D) 20 ohms

Answer: C

R = E/I; R = 120/7.5 = 16 ohms

2-6 An ohmmeter shows the resistance of a 240-volt electric heater to be 19.5 ohms. What current flows through this heater?

A) 10.5 amperes
B) 6.75 amperes
C) 3.2 amperes
D) 12.3 amperes

Answer: D

I = E/R; I = 240/19.5 = 12.3 amperes

2-7 An incandescent lamp has a resistance of 104 ohms when 2 amperes of current flow. What is the voltage?

A) 240 volts
B) 208 volts
C) 120 volts
D) 12 volts

Answer: B

E = IR; 2 × 104 = 208 volts

2-8 What current is drawn by a 277-volt fluorescent lamp with 8 ohms reactance?

A) 20.52 amperes
B) 60.75 amperes
C) 34.62 amperes
D) 10.30 amperes

Answer: C

$I = E/X_L$; I = 277/8 = 34.62 amperes

2-9 If the impedance Z of the circuit impedance triangle in Figure 2-3 is 5 ohms and angle A is 30°, what is the value of the inductive reactance (X_L)?

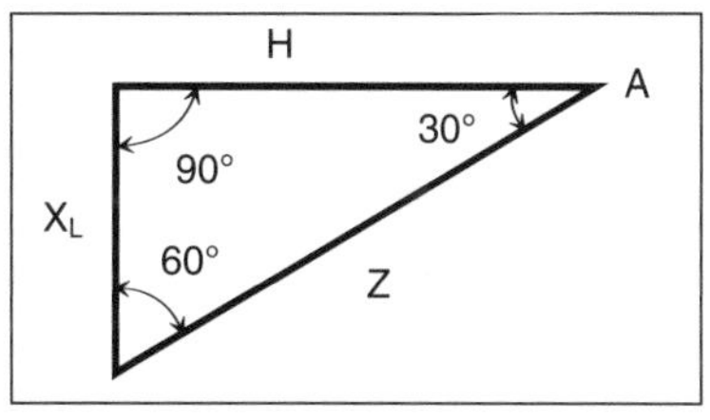

Figure 2-3: Impedance triangle

A) 2.5 ohms
B) 6.75 ohms
C) 3.2 ohms
D) 12.3 ohms

Answer: A

$$sin\ angle\ A = \frac{side\ opposite}{H} = sin\ 30(.500000) = \frac{X_L}{5}; X_L = .500000 \times 5 = 2.5\ ohms$$

2-10 What is the minimum general lighting load permitted by the *NEC* in a 4500 square foot warehouse?

A) 1005 volt-amperes
B) 1125 volt-amperes
C) 1750 volt-amperes
D) 1243 volt-amperes

Answer: B

Total volt-amperes = total square feet × 0.25 volt-amperes per square foot = 4500 × 0.25 = 1125 volt-amperes. NEC Table 220.12.

2-11 What is the minimum general lighting load permitted in a 1500 square foot single-family dwelling as specified in Table 220.12 of the 2005 *NEC*?

A) 4500 volt-amperes
B) 3500 volt-amperes
C) 7350 volt-amperes
D) 2130 volt-amperes

Answer A

1500 × 3 = 4500 volt-amperes. Note: this answer does not take into consideration the circuits required for small-appliance and laundry loads; these must be calculated separately.

2-12 The voltage per turn for a transformer is 1.25 volts. What is the voltage of the transformer if it has 192 turns?

A) 12 volts
B) 240 volts
C) 120 volts
D) 24 volts

Answer: B

192 × 1.25 = 240 volts

2-13 A 50 kVA, three-phase, delta-to-wye connected, 480/120-208 volt transformer is used to supply a lighting load. What is the rated line current on the primary side?

A) 45.54 amperes
B) 50.41 amperes
C) 60.21 amperes
D) 124.7 amperes

Answer: C

$$\frac{50000}{480\times\sqrt{3}}=\frac{50000}{480\times1.73}=\frac{50000}{830.4}=60.21\ \textit{amperes}$$

2-14 A 50 kVA, three-phase, delta-to-wye connected 480/120-208 volt transformer used to supply a balanced lighting load will have what line-current rating on the secondary?

A) 60.21 amperes
B) 138.95 amperes
C) 152.73 amperes
D) 176.54 amperes

Answer: B

$$\frac{50000}{208\times\sqrt{3}}=\frac{50000}{208\times1.73}=\frac{50000}{359.84}=138.95\ \textit{amperes}$$

2-15 A 480/277 volt, Y-connected transformer is used to supply a balanced 277-volt, single-phase lighting load of 40,000 watts. What size transformer (kVA) should be used?

A) 10 kVA

B) 20 kVA

C) 40 kVA

D) 50 kVA

Answer: C

A watt is a watt, is a watt....

2-16 What is the maximum allowable wattage that can be connected to a 240-volt, 20-ampere, single-phase circuit feeding residential electric heaters?

A) 3840 watts

B) 4000 watts

C) 4320 watts

D) 1920 watts

Answer: A

240 × 20 × 0.80 = 3840 watts. A circuit feeding electric heaters is considered continuous and must be reduced to 80% of its normal rating.

2-17 An electric oven is rated at 2000 watts when connected to a 460-volt circuit. What is the resistance of the heating element?

A) 50.3 ohms

B) 75.2 ohms

C) 105.8 ohms

D) 120.7 ohms

Answer: C

$$R = E^2 / W = \frac{460^2}{2000} = 105.8 \textit{ ohms}$$

2-18 What is the approximate wattage of a 240-volt motor with a resistance of 28 ohms?

A) 1780 watts

B) 2057 watts

C) 3640 watts

D) 1045 watts

Answer: B

$$\frac{240^2}{28} = 2057.14 \textit{ watts}$$

2-19 The minimum allowable number of 120-volt, 15-ampere, 2-wire lighting branch circuits required for a residence 70 feet by 30 feet are:

A) 2
B) 4
C) 3
D) 5

Answer: B

The area of the residence is first found (70 × 30 = 2100 square feet). Lighting branch circuits for a residence are based on 3 volt-amperes per square foot. Thus, 2100 × 3 = 6300 volt-amperes or watts. To find amperes, divide the voltage (120) into the volt-amperes; this equals 52.50 amperes. Since we are discussing 15-ampere circuits, 15 is divided into 52.50 = 3.5; the next higher "full" circuit being 4. Therefore, four 15-ampere branch circuits are required.

2-20 A voltage drop of 3 volts has been determined on a 120-volt circuit. What percent is the voltage drop?

A) 2%
B) 2.5%
C) 3%
D) 5%

Answer: B

3/120 = .025 or 2.5%

2-21 A voltage drop of 8.31 volts has been determined on a 277-volt lighting circuit. What percent is the voltage drop?

A) 2%
B) 2.5%
C) 3%
D) 4%

Answer: C

8.31/277 = .03 or 3%

2-22 A voltage drop of 7.2 volts is found on a 240-volt circuit. What percent is the voltage drop?

A) 1%
B) 2%
C) 3%
D) 4%

Answer: C

7.2/240 = .03 or 3%

2-23 A voltage drop of 19.2 volts has been determined on a 480-volt circuit. What percent is the voltage drop?

A) 2%

B) 2.5%

C) 4%

D) 5%

Answer: C

19.2/480 = .04 or 4%

2-24 If a 120-volt circuit feeds six 80-watt fluorescent lamps, what current is drawn if the power factor is 0.91?

A) 4.4 amperes

B) 3.2 amperes

C) 2.57 amperes

D) 4.7 amperes

Answer: A

$$6 \times 80 = \frac{480\ \textit{watts}}{120 \times 0.91} = 4.395\ \textit{or}\ 4.4\ \textit{amperes}$$

2-25 Calculate the minimum volt-ampere capacity of a 155000 square foot school.

A) 225 kVA

B) 155 kVA

C) 375 kVA

D) 465 kVA

Answer: D

NEC Table 220.3(A) gives the minimum volt-amperes for a school as 3 per square foot. 155000 × 3 = 465000 or 465 kVA.

2-26 A store building 50 feet x 60 feet uses fluorescent lighting fixtures to illuminate the interior. If each fixture is to illuminate 60 square feet, how many light fixtures will have to be installed?

A) 30

B) 40

C) 50

D) 60

Answer: C

50 × 60 = 3000 square feet/60 = 50 lighting fixtures

2-27 An electric heating element has an energized resistance of 30 ohms and is connected to a 120-volt circuit. How much current will flow in the circuit?

A) 2 amperes

B) 4 amperes

C) 6 amperes

D) 8 amperes

Answer: B

I = E/R; 120/30 = 4 amperes

2-28 How many watts of heat are being produced by the heating element in Question 2-27?

A) 550 watts

B) 480 watts

C) 225 watts

D) 400 watts

Answer: B

W = E × I; 120 × 4 = 480 watts

2-29 A 240-volt circuit has a current flow of 20 amperes. How much resistance (ohms) is connected to the circuit?

A) 3 ohms

B) 6 ohms

C) 9 ohms

D) 12 ohms

Answer: D

R = E/I; 240/20 = 12 ohms

2-30 An electric motor has an apparent resistance of 17 ohms. If a current of 7 amps is flowing through the motor, what is the connected voltage?

A) 118 volts

B) 119 volts

C) 120 volts

D) 121 volts

Answer: B

E = I × R; 7 × 17 = 119 volts

2-31 A 240-volt air-conditioning compressor has an apparent resistance of 8 ohms. How much current will flow in the circuit?

A) 10 amperes
B) 20 amperes
C) 30 amperes
D) 40 amperes

Answer: C

I = E/R; 240/8 = 30 amperes

2-32 How many watts of power are being used by the compressor in Question 2-31?

A) 5000 watts
B) 6500 watts
C) 7000 watts
D) 7200 watts

Answer: D

W = E × I; 240 × 30 = 7200 watts

2-33 A 4160-watt electric heater is connected to a 208-volt single-phase, 2-wire circuit. What is the current flow in the circuit?

A) 19 amperes
B) 22 amperes
C) 18 amperes
D) 20 amperes

Answer: D

I = W/E; 4160/208 = 20 amperes

2-34 What is the minimum size THW conductor allowed to carry the load in Question 2-33?

A) No. 14 AWG
B) No. 12 AWG
C) No. 10 AWG
D) No. 8 AWG

Answer: C

NEC Section 424.3(B) requires branch circuits feeding fixed electric space heating equipment to be sized not less than 125% of the total load on the circuit. Therefore, 20 × 1.25 = 25 amperes. No. 12 AWG THW wire is rated for 25 amperes, but the overcurrent protection cannot exceed 20 amperes. This circuit will require No. 10 AWG THW wire protected at 30 amperes.

2-35 A clamp-on ammeter has three turns of wire wrapped around the movable jaw. If the meter is indicating a current of 15 amperes, how much current is actually flowing in the circuit?

A) 5 amperes
B) 10 amperes
C) 25 amperes
D) 45 amperes

Answer: A

If three turns of wire are wrapped around the jaw of the ammeter, the primary winding now contains three turns instead of one, and the turns ratio of the transformer is changed. The ammeter will now indicate triple the amount of current that is actually in the circuit. Therefore 15/3 = 5 amperes of current actually in the circuit.

2-36 A dual-voltage three-phase motor draws a current of 45 amperes when connected to a 240-volt, three-phase circuit. How much current will the same motor draw if connected to a 480-volt, three-phase circuit?

A) 11.3
B) 22.5
C) 30.4
D) 45

Answer: B

Since the voltage doubles, the amperage will be half.

2-37 A synchronous motor has an excitation voltage of 125 volts with a maximum rotor current of 10 amperes. What is the resistance of the motor?

A) 10.1 ohms
B) 11.5 ohms
C) 12.5 ohms
D) 13.4 ohms

Answer: C

R = E/I; 125/10 = 12.5 ohms

2-38 The primary of a transformer is connected to 120-volts. There are 30 volts on the secondary with a resistance of 5 ohms. How much current will flow in the primary of the transformer?

A) .50 amperes

B) .75 amperes

C) 1.25 amperes

D) 1.5 amperes

Answer: D

$$I_S = \frac{E_S}{R_S}; I_S = \frac{30}{5} = 6 \textit{ amperes in secondary}$$

$$I_p = \frac{E_S \times I_S}{E_p}; I_p = \frac{30 \times 6}{120}; I_p = 1.5 \textit{ amperes in the primary}$$

2-39 What current flows in a 60 Hz 120-volt circuit with a capacitance of 10μf (10 microfarads)?

A) .452 amperes

B) .555 amperes

C) 1.12 amperes

D) .786 amperes

Answer: A

Find capacitive reactance: $X_c = \frac{1}{2 \times \pi \times F \times C} = \frac{1}{2 \times 3.1416 \times 60 \times .000010}; X_c = 265.25 \textit{ or } 265.3 \textit{ ohms}$

Now use the equation $I = E/X_c$ *to obtain 120/265.3 = .452 amperes*

2-40 What is the total resistance in a 2-wire circuit, 30 feet in length, utilizing No. 10 AWG THWN conductors?

A) .050 ohms

B) .060 ohms

C) .070 ohms

D) .080 ohms

Answer: B

Use equation $R = \frac{K \times L}{CM}; R = \frac{10.4 \times 60}{10,380} = .060$ *ohms. A length (L) of 60 feet is used because the 2-wire cable is 30 feet long. This means 30 feet of wire going to the load, and 30 feet returning to the source. No. 10 wire has an area of 10,380 circular mils.*

2-41 If the circuit in Question 2-40 is connected to a 240-volt source and feeds a load drawing 24 amperes, what is the amount of voltage drop?

A) 2.24 volts
B) 3.24 volts
C) 1.44 volts
D) 0.44 volts

Answer: C

$E = I \times R; 24 \times .060 = 1.44$ *volts*

2-42 Referring to Question 2-41 above, how many volts in the circuit actually reach the load?

A) 240 volts
B) 241.44 volts
C) 242.44 volts
D) 238.56 volts

Answer: D

240 *(volts)* $- 1.44$ *(volts)* $= 238.56$ *volts*

2-43 A circuit consisting of two No. 14 AWG conductors is run from the panelboard to a load 100 feet away. If the K factor is 10.4 and No. 14 wire is equal to 4107 circular mils, what is the total resistance of the circuit?

A) 50.6 ohms
B) .506 ohms
C) 506 ohms
D) 5.06 ohms

Answer: B

$$R = \frac{K \times L}{CM} = \frac{10.4 \times 200}{4107} = .506\ ohms$$

2-44 What is the voltage drop in the circuit in Question 2-43 if the current is 24 amperes at 240 volts?

A) 12.144 volts
B) 1.214 volts
C) .1214 volts
D) .0121 volts

Answer: A

$E = I \times R; E = 24 \times .506 = 12.144$ *volts*

2-45 How much voltage actually reaches the load in the circuit in Question 2-44?

A) 240 volts
B) 237.8 volts
C) 227.9 volts
D) 217.8 volts

Answer: C

240 − 12.144 = 227.856 or 227.9 volts

2-46 What is the percentage of voltage drop in the circuit described in Question 2-45?

A) .0843 or about 8%
B) .0721 or about 7%
C) .0607 or about 6%
D) .0506 or about 5%

Answer: D

12.144/240 = .0506, or 5%

2-47 What is the inductance reactance in a 120-volt, 60 Hz circuit connected to a coil with an inductance of .7 henrys?

A) 2.639 ohms
B) 26.39 ohms
C) 263.9 ohms
D) .2639 ohms

Answer: C

$X_L = 2\pi \times F \times L = 2 \times 3.1416 \times 60 \times .7 = 263.9$ *ohms*

2-48 What current flows in the circuit in Question 2-47?

A) .0455 amperes
B) .455 amperes
C) 4.55 amperes
D) 45.5 amperes

Answer: B

$I = E/X_L = 120/263.9 = .455$ *amperes*

2-49 What is the impedance (Z) in a circuit with 40 ohms resistance (R) and 30 ohms inductive reactance (X_L)?

A) 20 ohms

B) 30 ohms

C) 40 ohms

D) 50 ohms

Answer: D

$$Z=\sqrt{R^2+X_L^2}=\sqrt{40^2+30^2}=\sqrt{1600+900}=\sqrt{2500}=50\ ohms$$

2-50 What is the power factor of a circuit with 400 watts true power, 300 vars reactive power, and 500 watts (volt-amperes) of apparent power?

A) .5

B) .6

C) .7

D) .8

Answer: D

PF = W/VA = 400/500 = .8 or 80%, when PF = power factor; W = true power in watts, and VA = apparent power in volt-amperes.

2-51 What is the power factor of a 120-volt circuit feeding an AC induction motor drawing 10 amperes of current with a true power of 720 watts and an apparent power of 1200 volt-amperes?

A) .2

B) .4

C) .6

D) .8

Answer: C

PF = W/VA = 720/1200 = .6 or 60%

2-52 What is the reactive power in the circuit in Question 2-51?

A) 960 vars

B) 1150 vars

C) 1225 vars

D) 1300 vars

Answer: A

$$Vars=\sqrt{VA^2-W^2}=\sqrt{1200^2-720^2}=\sqrt{1,440,000-518,400}=\sqrt{921,600}=960\ vars$$

2-53 What is the capacitive reactance of the circuit in Question 2-52?

A) 10 ohms

B) 15 ohms

C) 20 ohms

D) 25 ohms

Answer: B

$X_C = E^2/vars = 120^2/960 = 15\ ohms$

2-54 What amount of capacitance is needed to correct the power factor of the circuit in Question 2-52?

A) .01768 farads

B) .001768 farads

C) .0001768 farads

D) .00001768 farads

Answer: C

$$C = \frac{1}{2 \times \pi \times F \times X_c} = \frac{1}{2 \times 3.1416 \times 60 \times 15} = \frac{1}{5654.88} = .0001768\ farads$$

2-55 What is the total resistance of a series circuit with resistors of 5 ohms, 3 ohms, 12 ohms, and 9 ohms?

A) 20 ohms

B) 21 ohms

C) 29 ohms

D) 23 ohms

Answer: C

$R_t = R_1 + R_2 + R_3 + R_4 = 5 + 3 + 12 + 9 = 29\ ohms$

2-56 What is the resistance of a parallel circuit with resistors of 30 ohms and 50 ohms?

A) 22.50 ohms

B) 18.75 ohms

C) 6.85 ohms

D) 10.55 ohms

Answer: B

$$R_t = \frac{R_1 \times R_2}{R_1 + R_2} = \frac{30 \times 50}{30 + 50} = 18.75\ ohms$$

2-57 If the two circuits in Questions 2-55 and 2-56 were combined forming a series-parallel circuit, what would be the total resistance?

A) 47.75 ohms
B) 52.25 ohms
C) 10.25 ohms
D) 18.75 ohms

Answer: A

The total resistance in the parallel circuit is found first which equals 18.75 ohms. The total resistance in the series circuit is 29 ohms. These figures added together equal 47.75 ohms. Once the total resistance in the parallel circuit is found, it may then be treated as a series circuit, and the total resistance in a series circuit may be found by adding all resistances together. The sum of these is the total resistance in the circuit.

2-58 What is the total resistance of a series circuit with four Christmas lights connected, each of which has a resistance of 3 ohms?

A) 2 ohms
B) 4 ohms
C) 6 ohms
D) 12 ohms

Answer: D

3 ohms x 4 = 12 ohms

2-59 A 12-volt battery supplies a total load of 10 amperes. What would the amperage be if the battery voltage dropped 10% and the load resistance remained the same?

A) 1.2 amperes
B) 9 amperes
C) 10.5 amperes
D) 12.5 amperes

Answer: B

To find the resistance of the load: $R = E/I$. 12/10 = 1.2 ohms. To find the voltage when reduced by 10%: 10% of 12 volts = 1.2 volts. 12 volts − 1.2 volts = 10.8 volts. To find the current: $I = E/R$. 10.8/1.2 = 9 amperes.

2-60 A department store is illuminated with 215 fluorescent lighting fixtures and connected to a 120-volt supply. Each lighting unit draws 2.2 amperes. How many 20-ampere branch circuits are necessary to feed these fixtures if each branch circuit must not exceed 80% of the branch circuit rating?

A) 16
B) 18
C) 20
D) 30

Answer: D

215 fixtures × 2.2 amperes = 473 amperes, total load. A 20-ampere branch circuit loaded to 80% of its capacity = 20 × .80 = 16 amperes. 473 amperes (total load) / 16 = 29.56 or 30 circuits.

2-61 If the lighting units in Question 2-60 have a power factor of 0.8, what is the total power of the lighting load?

A) 41.4 kW
B) 45.4 kW
C) 44.5 kW
D) 48.5 kW

Answer: B

Power = volts × amperes × power factor. 120 volts × 473 amperes × .8 = 45,408 watts or 45.4 kW

2-62 What would the electric bill be for the lighting load in Question 2-61 for an 8-hour day if the energy costs 10 cents per kWh?

A) $31.45
B) $35.67
C) $36.32
D) $38.40

Answer: C

Cost of power = Total power × cost of power × time/1000; 45,408 × .10 x 8/1000 = $36.32

2-63 A warehouse requires 60 kW of general illumination using 150 watt incandescent lamps connected to a 120-volt source. What is the current drawn by each lamp?

A) 1.50 amperes
B) 1.25 amperes
C) 1 ampere
D) .75 ampere

Answer: B

I = P/E; 150/120 = 1.25 amperes

2-64 How many 20-ampere branch circuits are required to feed the lamps in Question 2-63 if the circuits are to be kept at 80% of their maximum rating?

A) 30

B) 31

C) 32

D) 33

Answer: C

20 × .8 = 16 amperes per circuit. 60,000 VA/120 = 500 amperes
500 amperes/16 amperes per circuit = 31.25 or 32 circuits

2-65 What number of 150-watt lamps are required to total 60,000 watts?

A) 100

B) 200

C) 300

D) 400

Answer: D

60,000/150 = 400

2-66 An apartment complex consists of 20 apartments, each with an area of 900 square feet. Each is also equipped with an 11 kW electric range. The power supply is 120/240-volt, single-phase, three-wire. The *NEC* calls for a minimum lighting load of 3 volt-amperes per square foot. What is the total lighting and appliance load for each apartment?

A) 10,945 volt-amperes

B) 11,945 volt-amperes

C) 12,945 volt-amperes

D) 13,945 volt-amperes

Answer: B

900 square feet × 3 volt-amperes = 2700 volt-amperes; two 1500-volt-ampere small appliance circuits = 3000 volt-amperes. Total lighting and small appliance load = 5700 volt-amperes. The first 3000 volt-amperes must be rated at 100%. A demand factor of 35% may be applied to the remaining 2700 volt-amperes. 2700 × .35 = 945 volt-amperes for a total lighting and small appliance load of 3945 volt-amperes. A demand factor may be applied to the 11 kW electric range and rated at 8 kW or 8000 volt-amperes. Therefore, the total calculated load for each apartment is 3945 + 8000 = 11,945 volt-amperes. NEC Tables 220.12, 220.42 and 220.55.

2-67 What are the total current requirements for each apartment in Question 2-66?

A) 49.77 amperes
B) 50.77 amperes
C) 51.77 amperes
D) 52.77 amperes

Answer: A

$$I = \frac{P}{E} = \frac{11{,}945}{240} = 49.77 \textit{ amperes}$$

2-68 What is the total calculated load for the building in Question 2-66?

A) 70,000 volt-amperes
B) 72,000 volt-amperes
C) 75,000 volt-amperes
D) 77,000 volt-amperes

Answer: D

20 apartments × 5700 volt-amperes = 114,000 volt-amperes. The first 3000 volt-amperes must be rated at 100%. The remaining 111,000 volt-amperes may be derated at 35% for a total of 39,000 volt-amperes. This results in a net computed load of 42,000 volt-amperes without electric ranges. The demand load for 20 ranges according to NEC Table 220.55 = 35 kW. Therefore, 42,000 volt-amperes + 35,000 volt-amperes = 77,000 volt-amperes.

2-69 What size electric service is required for the apartment building in Question 2-68?

A) 320.83 amperes
B) 420.83 amperes
C) 520.83 amperes
D) 620.83 amperes

Answer: A

77,000 volt-amperes/240 volts = 320.83 amperes

2-70 A 60 kW incandescent lighting load is connected to a 120/208-volt, three-phase, 4-wire electric service. The load is balanced. What is the total current consumed by this load?

A) 169.50 amperes
B) 160.25 amperes
C) 166.74 amperes
D) 169.24 amperes

Answer: C

$$\frac{60{,}000}{\sqrt{3} \times 208} = \frac{60{,}000}{1.73 \times 208} = 166.74 \textit{ amperes}$$

2-71 An industrial building is illuminated with 200 277-volt fluorescent lighting fixtures connected to a 277/480-volt, three-phase, 4-wire, Y-connected supply. If each fixture draws 1.2 amperes at a power factor of .8, what is the total load in amperes per single phase?

A) 120 amperes

B) 480 amperes

C) 220 amperes

D) 240 amperes

Answer: D

200 × 1.2= 240 amperes

2-72 On a 277/480 volt three-phase supply with a balance load of 240 amperes, what is the current per phase?

A) 70 amperes

B) 80 amperes

C) 90 amperes

D) 100 amperes

Answer: B

240 amperes / 3 = 80 amperes. The use of a 277/480 volt supply permits 1/3 of the load to be connected to each phase on a perfectly balanced load.

2-73 A group of electric furnaces used in a heat-treating plant require 60 kW of power at 240 volts, single-phase. What is the total amperage of these furnaces at unity power factor?

A) 100 amperes

B) 150 amperes

C) 200 amperes

D) 250 amperes

Answer: D

I = P/E; 60,000/240 = 250 amperes

2-74 How many 15-ampere lighting circuits are necessary in a mobile home 12 feet by 65 feet?

A) Two 15-ampere circuits

B) One 15-ampere circuit

C) Six 15-ampere circuits

D) Four 15-ampere circuits

Answer: A

3 × L × W/120 × 15 = No. of circuits required. NEC Section 550.12(A) requires that 3 volt-amperes (watts) per square foot be used for lighting circuits in mobile homes. Therefore, 3 × 12 × 65/(120 × 15) = 1.3 or two 15-ampere circuits are required.

Chapter 3

Branch Circuits and Feeders

The *NEC* defines "branch circuit" as the circuit conductors between the final overcurrent device protecting the circuit and the outlet(s).

The *NEC* defines "feeder" as all circuit conductors between the service equipment or the source of a separately derived system and the final branch-circuit overcurrent device.

Electric power is delivered to panelboard locations by feeder conductors extending from the main service-entrance equipment to the branch-circuit panelboards or load centers. The feeder conductors must be of sufficient size to meet the requirements of *NEC* Articles 220 and 310, and are calculated roughly on a minimum basis of allowing for the connecting lighting load or a certain number of watts per square foot for lighting, plus the power equipment or appliance load requirements. In some cases, a demand factor is allowed for the connected power load.

On larger electrical installations, distribution centers are established to which point large feeder conductors are run from the main service equipment. Subfeeders are then run from the distribution center to the various lighting and power panelboards.

The diagram in Figure 3-1 on the next page shows the service conductors, feeders to subpanels, and branch circuits.

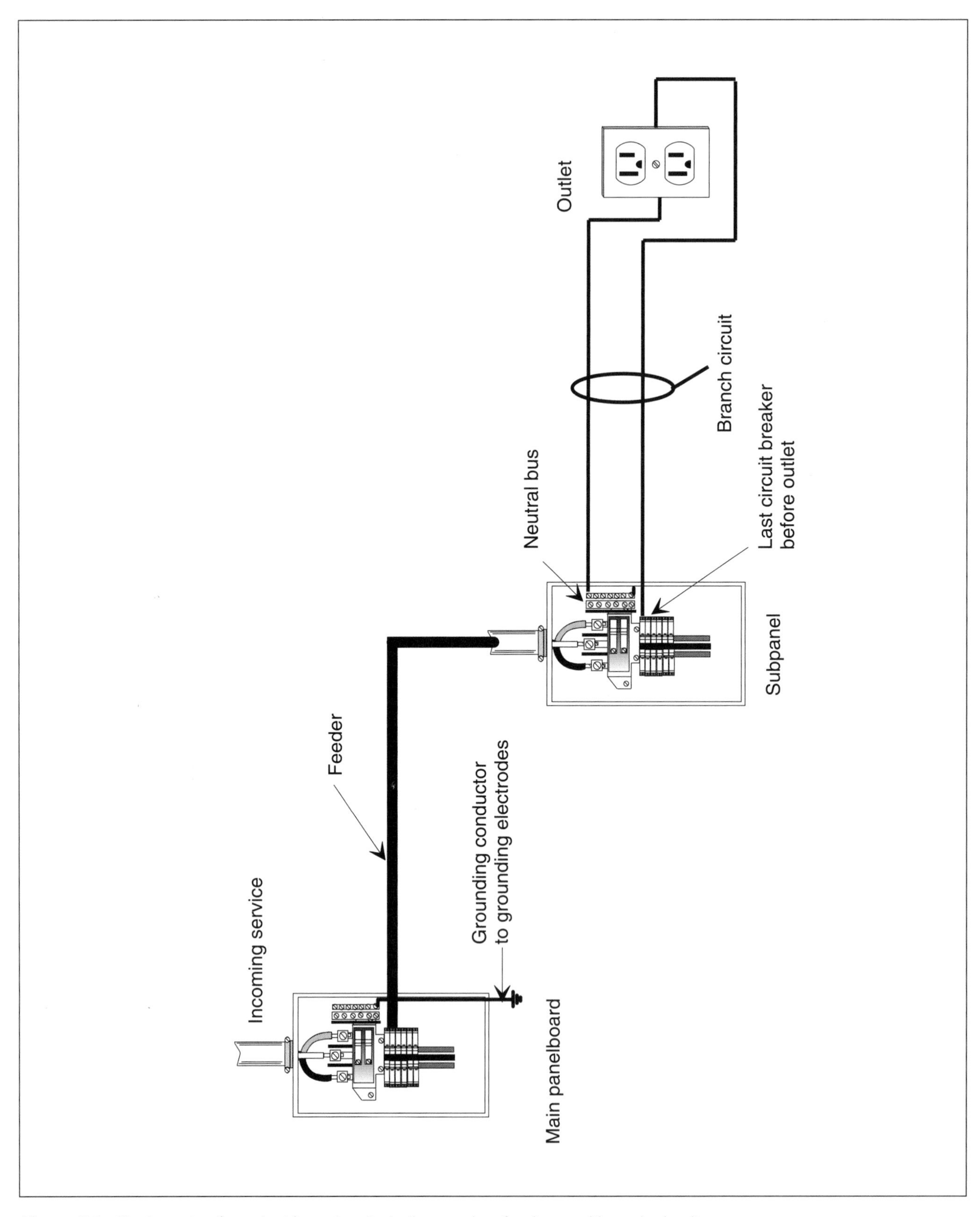

Figure 3-1: Basic parts of an electric system including service, feeders and branch circuits

3-1 The cross-sectional area of three or more THW conductors in a conduit must not exceed what percent of the cross-sectional area of the conduit?

A) 30%

B) 40%

C) 50%

D) 60%

Answer: B

NEC Chapter 9, Table 1. See Figure 3-2.

Figure 3-2: Sectional view of raceway with conductor fill

3-2 What is the minimum size THW copper conductor that may be used on a 30-ampere branch circuit?

A) No. 14 AWG

B) No. 12 AWG

C) No. 10 AWG

D) No. 8 AWG

Answer: C

Conductor current-carrying capacity is based on the cross-sectional area of the conductor, the type of metal used to make up the conductor, the type of insulation, the number of conductors in a raceway, and ambient temperatures in which the conductors will operate. NEC Table 310.16 and 240.4(D).

3-3 Heavy-duty type lampholders are required on branch-circuits having a rating in excess of:

A) 20 amperes

B) 30 amperes

C) 40 amperes

D) 15 amperes

Answer: A

Lampholders connected to a branch circuit having a rating in excess of 20 amperes must be of the heavy-duty type. Heavy-duty lampholders must be rated no less than 660 watts if of the admedium type and not less than 750 watts if of any other type. NEC Section 210.21(A).

3-4 What is the maximum *continuous* load (in amperes) that can be used on a 240-volt, 20-ampere circuit using two No. 12 AWG conductors with THW insulation?

A) 12 amperes

B) 15 amperes

C) 16 amperes

D) 20 amperes

Answer: C

Circuits handling a continuous load must have conductors rated 125% above the connected load. NEC Section 210.20(A). Since a No. 12 conductor is rated at 20 amperes for normal loads, the maximum current permitted for continuous loads is 16 amperes.

$$\frac{20}{125\%} = 16 \text{ amps}$$

3-5 A 30-ampere branch circuit may supply fixed lighting units in:

A) A dwelling

B) Any occupancy other than dwelling units when heavy-duty lampholders are used

C) Any type of occupancy

D) Any occupancy other than dwelling units when standard-duty lampholders are used

Answer: B

A 30-ampere branch circuit shall be permitted to supply fixed lighting units with heavy-duty lampholders in other than dwelling units. NEC Section 210.23(B).

3-6 Branch circuits in two-family or multifamily dwellings required for the purpose of central lighting, alarm, signal, communications, or other needs for public or commercial areas shall:

A) Be rated over 30 amperes for any application or purpose

B) Terminate in one of the dwelling unit's panelboards

C) Be rated over 50 amperes

D) Not be supplied from a dwelling unit's panelboard

Answer: D

This restriction is specified in NEC Section 210.25. It means that "common-use" electrical circuits as described above must be fed from the "house" panelboard and not from one of the tenant's panelboards, or one used for light and power for an individual unit.

3-7 EMT must be supported within what distance from an outlet box, junction box, device box, cabinet, conduit body or other terminations?

A) 3 feet

B) 4 feet

C) 5 feet

D) 6 feet

Answer: A

Different support distances are required for different types of raceway and cable. NEC Section 358.30(A).

3-8 Receptacles installed in guest rooms in hotels, motels and similar occupancies:

A) Are not subject to *NEC* regulations

B) Must be spaced according to *NEC* Section 210.52

C) May be located conveniently for permanent furniture layout

D) May be installed only if prior approval is given by the local inspection authority

Answer: C

NEC Section 210.52 must be used to determine the amount of receptacles installed. Section 210.60(B) allows the receptacles to be located conveniently for permanent furniture layout.

3-9 All 125-volt, single-phase, 15- or 20-ampere receptacle outlets installed in bathrooms of dwelling units always require:

A) Twist-lock receptacles

B) Ground-fault circuit-interrupter protection

C) Non-grounded receptacles

D) Waterproof receptacle covers

Answer: B

Many people have been electrocuted in bathrooms due to an electric appliance (radio, electric razor, etc.) falling into the sink or bathtub where a person is in contact with the water. A GFCI prevents many such accidents. NEC Section 210.8(A)(1).

3-10 The *NEC* allows wall-switch controlled receptacles in dwelling units in lieu of lighting outlets in all habitable rooms except:

A) Basement and attic

B) Living and family rooms

C) Bedrooms

D) Kitchen and bath

Answer: D

The NEC allows a switched receptacle in all rooms, in lieu of ceiling or wall-mounted lighting fixtures, except the kitchen and bathroom. NEC Section 210.70(A), Exception No. 1.

3-11 Insulated conductors of No. 6 or smaller intended for use as grounded conductors in a circuit must be either of the following colors:

A) Blue or black

B) Red or black

C) White, gray or have three continuous white stripes on other than green insulation

D) Orange or yellow

Answer: C

Insulated conductors, No. 6 AWG and smaller, intended for use as grounded (neutral) conductors or circuits, shall be identified by a continuous outer finish of a white or gray, or by three continuous white stripes along its entire length on other than green insulation. Conductors larger than No. 6 AWG may use white markers at points of termination. NEC Section 200.6(A) and 200.6(B).

3-12 Holes in wood studs for cables must be drilled in the center of the stud and a distance from the nearest edge of not less than:

A) 1/2 inch

B) 1 inch

C) 1 1/4 inches

D) 1 1/2 inches

Answer: C

NEC Section 300.4(A)(1). See Figure 3-3.

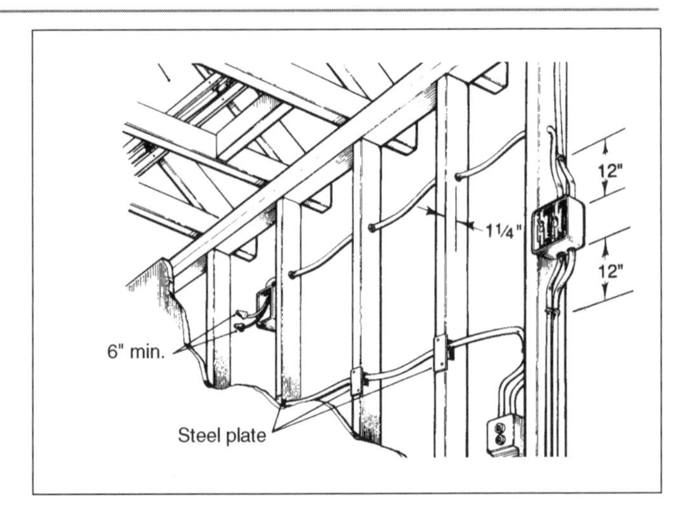

Figure 3-3: Type NM cable installed in wooden studs

3-13 Direct buried conductors and cables emerging from the ground shall be protected by an enclosure or raceway a distance above ground of:

A) 6 feet

B) 8 feet

C) 10 feet

D) 12 feet

Answer: B

Conductors emerging from the ground are subject to damage from lawn mowers, weed eaters, and the like. Therefore, they must be protected by some means. Metal conduit, such as EMT or rigid steel conduit are two approved means of protection. NEC Section 300.5(D).

3-14 Cable tray systems must *not* be used:

A) For power and control applications

B) For service-entrance systems

C) For signal cables

D) In hoistways

Answer: D

Wiring installations installed in hoistways are subject to physical damage. Cable trays, using open conductors, are not permitted in ducts, plenums and other air-handling spaces. NEC Section 392.4.

3-15 EMT shall be supported at least every:

A) 4 feet

B) 8 feet

C) 10 feet

D) 15 feet

Answer: C

EMT must be supported every 10 feet to prevent sag or collapse. NEC Section 358.30(A).

3-16 The total equivalent bend in rigid conduit between pull points (outlet to outlet, fitting to fitting, or outlet to fitting) shall not exceed:

A) 3 quarter bends (270°)

B) 4 quarter bends (360°)

C) 5 quarter bends (450°)

D) 6 quarter bends (540°)

Answer: B

For example, two 45° bends plus three 90° bends equals the maximum of 360° total. NEC Section 344.26. *See Figure 3-4 on the next page.*

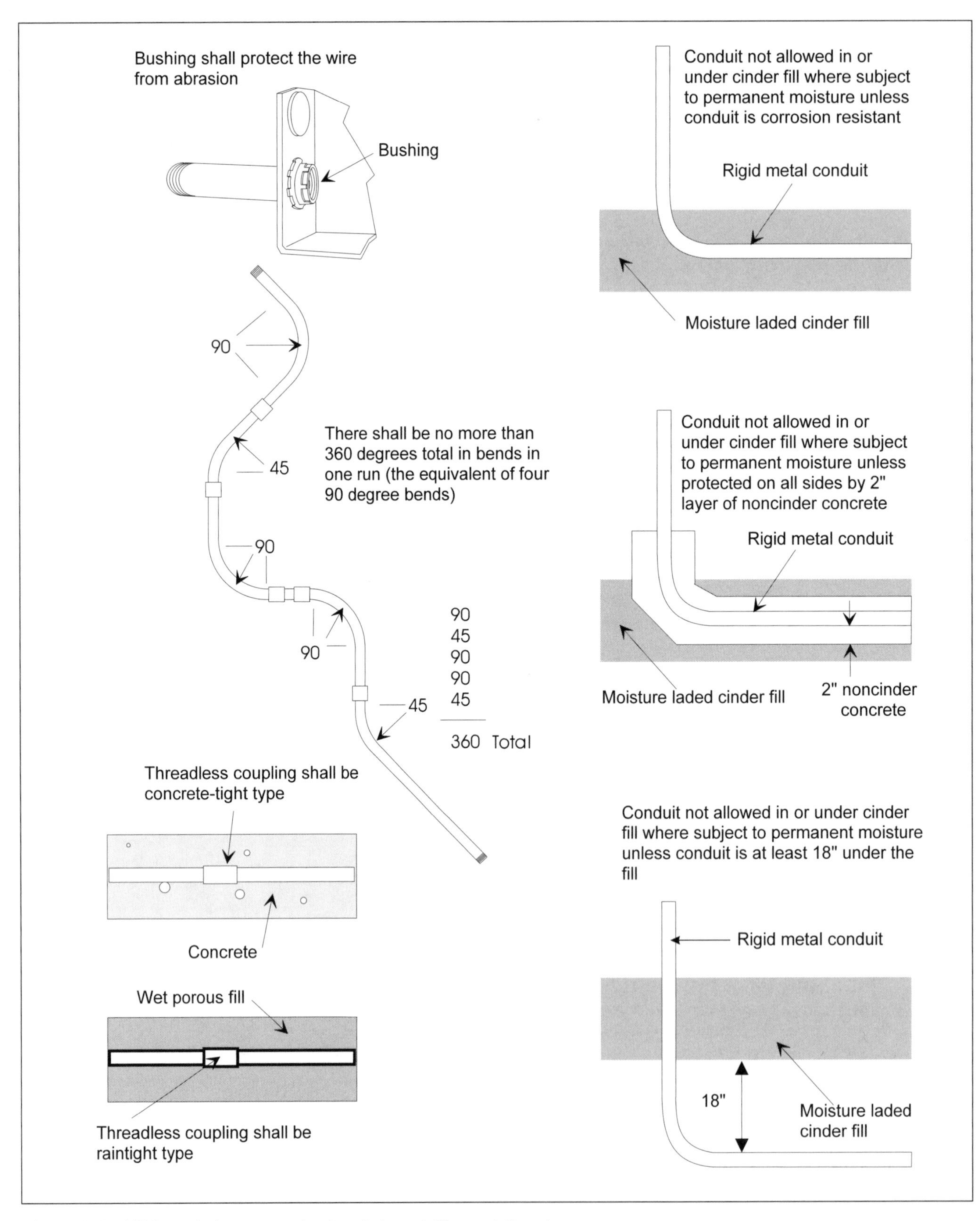

Figure 3-4: *NEC* regulations governing bends in metallic conduit systems

3-17 When a branch circuit feeds both noncontinuous and continuous loads, the circuit must be sized to accommodate 100% of the noncontinuous load plus what percent of the continuous load?

A) 110%
B) 115%
C) 120%
D) 125%

Answer: D

The rating of branch circuits serving continuous loads shall be not less than the noncontinuous load plus 125% of the continuous load. NEC Section 210.20(A).

3-18 The branch-circuit load for a single 12-kilowatt (12 kW) electric range in a dwelling is:

A) 8 kW
B) 9 kW
C) 10 kW
D) 11 kW

Answer: A

NEC Table 220.55 permits an electric range with a kW nameplate rating not exceeding 12 kW to be rated at 8 kW when sizing branch circuits and feeders.

3-19 A 20-ampere laundry circuit is always required in each:

A) Industrial wash room
B) Commercial utility room
C) Dwelling unit
D) Motel unit

Answer: C

In dwelling units, at least one 20 amp branch circuit shall be installed for the laundry. NEC Section 210.11(C)(2).

3-20 How many receptacles are required in a residential hallway 12 feet in length?

A) 1
B) 2
C) 3
D) 4

Answer: A

NEC Section 210.52(H) requires at least one receptacle outlet in hallways 10 feet or more in length.

3-21 How many receptacle outlets are required in a residential garage?

A) 4

B) 3

C) 2

D) 1

Answer: D

NEC Section 210.52(G) requires at least one receptacle outlet to be installed in each basement and in each attached garage, or in each detached garage with electric power. Also see NEC Sections 210.8(A)(2) and (A)(5).

3-22 A 125-volt single-phase, 15- or 20-ampere receptacle must be installed in locations for the servicing of heating, air conditioning and refrigeration equipment?

A) All apartment buildings

B) One- and two-family dwellings

C) Commercial buildings

D) All locations

Answer: D

Receptacle outlets are required for equipment in any location the equipment is present. The NEC has removed the exception for equipment on rooftops of one- and two-family dwellings. NEC Section 210.63.

3-23 The receptacle in Question 3-22, other than for one- and two-family dwelling units, must be installed within what distance from the equipment?

A) 25 feet

B) 50 feet

C) 75 feet

D) 100 feet

Answer: A

NEC Section 210.63 requires the receptacle to be on the same level and within 25 feet of the equipment. That is, if the equipment is located in, say, a crawl space, the receptacle must also be located in the crawl space. If the equipment is in the attic, the receptacle must be located there also.

3-24 On what side of the equipment disconnecting means must the receptacle in Question 3-22 *not* be connected?

A) On the line side

B) At the equipment controls

C) On the load side

D) At the main panelboard

Answer: C

NEC Section 210.63 states that the service receptacle must not be connected to the load side of the equipment disconnecting means.

3-25 How many lighting outlets must be installed at or near heating, air-conditioning, and refrigeration equipment in attics or crawl spaces?

A) 1

B) 2

C) 3

D) 4

Answer: A

NEC Sections 210.70(A) and (C) require at least one switched lighting outlet in these spaces.

3-26 How must the lighting outlet in Question 3-25 be controlled?

A) From the load side of the HVAC equipment

B) With a motor control

C) By a switch

D) With a 30-amp safety switch

Answer: C

NEC Section 210.70(C) calls for a lighting outlet containing a switch or controlled by a wall switch. (A porcelain-type lampholder with a pull-chain switch would meet the requirements.)

3-27 Where must the wall switch, as called for in Question 3-26, be located?

A) Mounted on the HVAC equipment

B) At the point of entry to the area

C) At the main disconnect

D) At least 5 feet from the equipment

Answer: B

NEC Sections 210.70(A) and (C) require the outlet or wall-mounted switch to be located at the usual point of entry.

3-28 Feeders containing a common neutral are permitted to supply how many sets of 3-wire feeders?

A) 1 or 2
B) 5 or 6
C) 3 or 4
D) 2 or 3

Answer: D

NEC Section 215.4(A) permits "two or three" sets of 3-wire feeders.

3-29 Feeders containing a common neutral are permitted to supply how many sets of 4-wire feeders?

A) 2
B) 3
C) 4
D) 5

Answer: A

NEC Section 215.4(A) permits two sets of 4-wire feeders.

3-30 Feeders containing a common neutral are permitted to supply how many sets of 5-wire feeders?

A) 1
B) 2
C) 3
D) 4

Answer: B

NEC Section 215.4(A) permits a common neutral to supply two sets of 5-wire feeders.

3-31 Service conductors run above the top level of a window shall be permitted to be less than the _____ requirement.

A) 3-foot
B) 6-foot
C) 8-foot
D) 10-foot

Answer: A

NEC Section 225.19(D)(1), Exception permits these conductors to be less than the 3-foot requirement.

3-32 Which one of the following voltages is not standard?

A) 120/275

B) 480Y/277

C) 120/240

D) 600Y/347

Answer: A

NEC Section 220.5(A) lists the common voltages as 120, 120/240, 208Y/120, 240, 347, 480Y/277, 480, 600Y/347 and 600.

3-33 Where a feeder conductor supplies continuous loads, the rating of the overcurrent device shall not be less than the noncontinuous load plus what percent of the continuous load?

A) 100%

B) 110%

C) 125%

D) 150%

Answer: C

NEC Section 215.2(A)(1) requires 125%.

3-34 What is one exception to the answer in Question 3-33?

A) Where the overcurrent devices are listed for operation at 100%

B) Where the overcurrent devices are listed at 75%

C) Where the overcurrent devices are listed at 80%

D) Where no overcurrent devices are provided

Answer: A

NEC Section 215.2(A)(1), Exception permits the use of a lesser percent when the assembly, including the overcurrent devices, is listed for operation at 100 percent of its rating.

3-35 What is the minimum load per each linear foot that must be allowed for show-window lighting?

A) 100 volt-amperes

B) 200 volt-amperes

C) 300 volt-amperes

D) 400 volt-amperes

Answer: B

NEC Section 220.14(G) requires a minimum of 200 volt-amperes be provided for each linear foot of show window, measured horizontally along its base.

3-36 What percent of fixed electric space heating loads must be used in calculating the total connected load on a branch circuit or feeder?

A) 100%
B) 125%
C) 150%
D) 200%

Answer: A

NEC Section 220.51 requires that fixed electric space heating be computed at 100% of the total connected load.

3-37 If four 12 kW electric ranges are used in a four-unit apartment complex, what is the total kW rating that must be used for all of these ranges in calculating the electric service?

A) 17 kW
B) 18 kW
C) 20 kW
D) 48 kW

Answer: A

NEC Table 220.55 allows a demand factor for four electric ranges of 17 kW provided their nameplate ratings do not exceed 12 kW each.

3-38 If the apartment complex in Question 3-37 is expanded to include a total of 40 units with a 12 kW range in each, what is the total kW rating for all the ranges that must be used in calculating the electric service?

A) 45 kW
B) 50 kW
C) 55 kW
D) 60 kW

Answer: C

When 31 to 40 electric ranges are used in an apartment complex, and the maximum kW rating of each is 12 kW, NEC Table 220.55 allows a calculation figure of 15 kW plus 1 kW for each range. Therefore, 15 kW + 40 kW = 55 kW.

3-39 What demand factor is allowed for a commercial kitchen with one electric appliance?

A) 65%
B) 70%
C) 80%
D) 100%

Answer: D

NEC Table 220.56 requires that a commercial kitchen with only one appliance be rated at 100%. There must be three or more appliances before a reduction is allowed.

3-40 When outside overhead wiring is used, with no messenger cable, what is the minimum size copper wire allowed for spans up to 50 feet?

A) No. 6 AWG
B) No. 8 AWG
C) No. 10 AWG
D) No. 12 AWG

Answer: C

NEC Section 225.6(A) restricts the wire size to No. 10 AWG copper for spans up to 50 feet on installations of 600 volts or less.

3-41 When outside overhead wiring is used, with no messenger cable, what is the minimum size aluminum wire allowed for spans up to 50 feet?

A) No. 6 AWG
B) No. 8 AWG
C) No. 10 AWG
D) No. 12 AWG

Answer: B

NEC Section 225.6(A) requires a minimum aluminum wire size of No. 8 AWG on installations of 600 volts or less.

3-42 When the voltage exceeds 600 volts, what is the minimum copper wire size allowed for outside overhead wiring?

A) No. 6 AWG
B) No. 8 AWG
C) No. 10 AWG
D) No. 12 AWG

Answer: A

NEC Section 225.6(A)(2) restricts the wire size to No. 6 AWG copper when the wires carry more than 600 volts.

3-43 Where outdoor pendant-type lampholders have terminals that puncture conductor insulation to make contact with the conductor, what type of wire must the conductor be?

A) Copper-plated solid steel
B) Copper-clad solid aluminum
C) Stranded
D) Solid copper

Answer: C

NEC Section 225.24 requires that this type of lampholder be connected only to conductors of the stranded type.

3-44 What is the minimum size conductor allowed for festoon lighting?

A) No. 8 AWG
B) No. 10 AWG
C) No. 12 AWG
D) No. 14 AWG

Answer: C

NEC Section 225.6(B) permits overhead conductors for festoon lighting to be as small as No. 12 AWG. However, spans exceeding 40 feet must be provided with a messenger wire.

3-45 Receptacles installed on 15- and 20-ampere branch circuits must always be of what type?

A) Duplex receptacles
B) Grounding type
C) Twist-lock receptacles
D) Three-prong type

Answer: B

NEC Section 406.3(A) requires that all receptacles installed on 15- and 20-ampere circuits be of the grounding type.

3-46 What is the maximum voltage allowed to supply listed electric-discharge lighting in residences, hotels, motels and similar occupancies?

A) 120 volts
B) 240 volts
C) 277 volts
D) 480 volts

Answer: A

NEC Section 210.6(A) restricts the voltage to 120 volts for lighting in occupancies.

3-47 Receptacles and cord connectors having grounding contacts must have those contacts effectively:

A) Energized

B) Grounded

C) Polished and free from corrosion

D) Checked monthly

Answer: B

NEC Section 406.3(B) requires that such receptacles and connectors must be effectively grounded, except when these receptacles and cord connectors are mounted on portable or vehicle-mounted generators in accordance with NEC Section 250.34.

3-48 The grounding contacts of receptacles and cord connectors must be grounded by:

A) Connection to the equipment grounding conductor

B) Aluminum ground clips

C) Aluminum conductors

D) Aluminum ground clamps

Answer: A

NEC Section 406.3(C) requires that these contacts shall be grounded by connection to the equipment grounding conductor of the circuit supplying the receptacle or cord connector.

3-49 When existing nongrounding types of receptacles need replacing, what type of receptacle must be used?

A) Receptacles rated for at least 15 amperes

B) Receptacles rated for at least 20 amperes

C) Twist-lock receptacles

D) Grounding-type receptacles

Answer: D

NEC Section 406.3(D) requires that existing nongrounding types of receptacles be replaced with the grounding type and connected to a grounding conductor. Where a grounding means does not exist in the receptacle enclosure, either a nongrounding or a ground-fault circuit-interrupter type of receptacle must be used.

3-50 When receptacles are connected to circuits having different voltages, frequencies, or types of current (ac or dc) on the same premises, what precautions must be taken?

A) One type must have a means of disconnection from the circuit when another type is used

B) Each receptacle must be tagged for the use intended

C) None of the receptacles are allowed to be used without proper supervision

D) The attachment plugs must be of a design so they are not interchangeable

Answer: D

NEC Section 406.3(F) requires that attachment plugs used on such circuits not be interchangeable.

3-51 What location requires all 15- and 20-ampere receptacles to be protected with a ground-fault circuit-interrupter?

A) Those installed within 6 feet of a residential kitchen sink

B) Those installed in hotel and motel bathrooms, rooftops and kitchens

C) Those installed in residential bathrooms

D) All of these areas

Answer: D

NEC Section 210.8 requires that all of these areas have receptacles protected with a GFCI.

3-52 Which one of the following is not allowed to supply branch circuits?

A) A 277-volt circuit

B) A 120-volt circuit

C) Autotransformer if the circuit is without a grounded conductor of the required type

D) A buck-and-boost transformer

Answer: C

NEC Section 210.9 does not allow branch circuits to be derived from autotransformers unless the circuit supplied has a grounded conductor that is electrically connected to a grounded conductor of the system supplying the autotransformer.

3-53 According to the *NEC*, as applied to electrical wiring systems, what constitutes a bathroom?

A) An area with shower curtains

B) An area with a sink

C) An area with a basin and either a toilet, tub or shower

D) An area that contains a water faucet

Answer: C

NEC Article 100 defines a bathroom as an area including a basin with one or more of the following: a toilet, a tub, or a shower. See Figure 3-5.

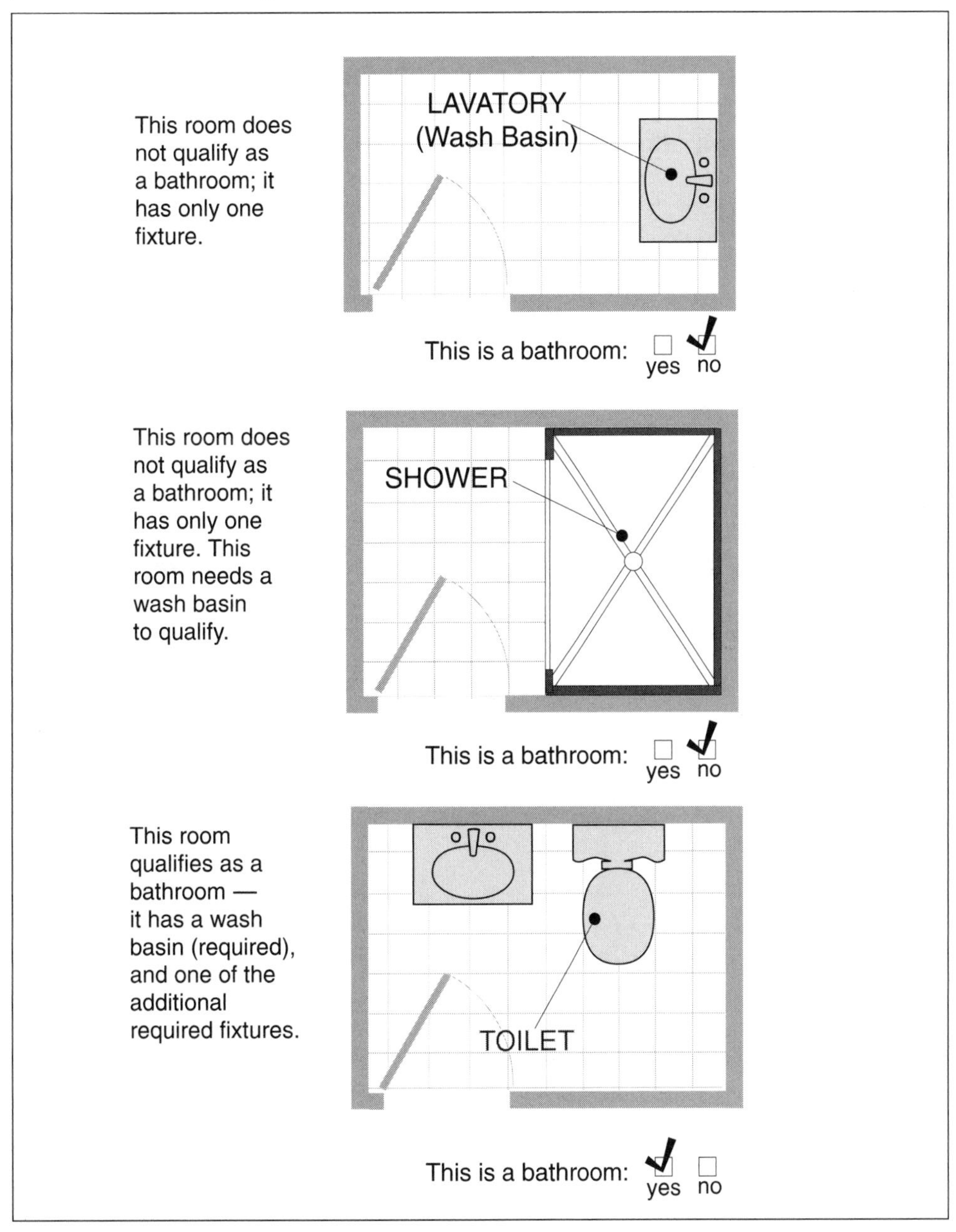

Figure 3-5: *NEC* definition of bathroom

3-54 What is the maximum load allowed by the *NEC* for cord- or plug-connected load to a 15-ampere receptacle?

A) 10 amperes
B) 12 amperes
C) 14 amperes
D) 16 amperes

Answer: B

NEC Table 210.21(B)(2) limits the load to 12 amperes when connected to either a 15 or 20 ampere circuit.

3-55 What is the maximum load allowed by the *NEC* for cord- or plug-connected load to a 20-ampere receptacle?

A) 10 amperes
B) 12 amperes
C) 14 amperes
D) 16 amperes

Answer: D

NEC Table 210.21(B)(2) stipulates 16 amperes as maximum for a 20-ampere receptacle.

3-56 What is the minimum rating allowed for a single receptacle connected to a branch circuit?

A) Not less than that of the branch circuit
B) Not less than 85% of the branch circuit rating
C) Not less than 90% of the branch circuit rating
D) Not less than 70% of the branch circuit rating

Answer: A

NEC Section 210.21(B) requires that a single receptacle have a rating no less than the branch circuit.

3-57 The minimum branch-circuit rating for household electric ranges with a rating of 8.75 kW or more is:

A) 30 amperes
B) 40 amperes
C) 50 amperes
D) 60 amperes

Answer: B

NEC Section 210.19(A)(3) requires that branch circuits supplying electric ranges with 8.75 kW or more be rated no less than 40 amperes.

3-58 What is the minimum rating for heavy-duty lampholders if not of the admedium type?

A) 550 watts

B) 660 watts

C) 750 watts

D) 770 watts

Answer: C

NEC Section 210.21(A) requires heavy-duty lampholders to have a rating of not less than 660 watts if of the admedium type and not less than 750 watts if of any other type.

3-59 Temporary wiring used on a construction project must be removed:

A) Two weeks before the completion of the project

B) Within six months after the completion of the project

C) One week before completion of the construction

D) Immediately upon completion of the construction

Answer: D

NEC Section 590.3(D) requires that temporary service be removed immediately upon completion of construction or purpose for which the wiring was installed. The actual time permitted will vary from jurisdiction to jurisdiction, but usually the temporary service is disconnected at the same time the permanent service is connected.

3-60 All temporary 125-volt, single-phase, 15-, 20-, and 30-ampere receptacle outlets must be provided with:

A) Twist-lock plugs

B) Four-prong plugs

C) Ground-fault circuit-interrupters

D) Five-prong plugs

Answer: C

NEC Section 590.6(A) requires all temporary receptacles to be protected with GFCIs for the protection of workers on the job.

3-61 When installed in a raceway, what size conductors must be stranded?

A) No. 12 AWG and larger

B) No. 8 AWG and larger

C) No. 10 AWG and larger

D) No. 2 AWG and smaller

Answer: B

NEC Section 310.3 requires that all No. 8 AWG and larger conductors be stranded to facilitate pulling them through the raceway system and for working in panelboards, pull boxes, and the like.

3-62 What is the minimum size conductor that may be connected in parallel?

A) 1/0 kcmil

B) 2/0 kcmil

C) 3/0 kcmil

D) 4/0 kcmil

Answer: A

NEC Section 310.4 allows aluminum, copper-clad aluminum, or copper conductors, 1/0 or larger, comprising each phase, neutral or grounded circuit conductor, to be connected in parallel.

3-63 When conductors are paralleled, they are:

A) Joined their full length

B) Electrically separated at both ends to form two separate conductors

C) Electrically joined at both ends to form a single conductor

D) Electrically connected at one end only

Answer: C

The ends of two or more conductors are connected to the same lug or terminal at each end of the conductor runs. NEC Section 310.4.

3-64 AFCIs must be installed on all branch circuits supplying?

A) Outdoor receptacles

B) Kitchen areas

C) Bathrooms

D) Bedrooms

Answer: D

Section 210.12(B) requires that all circuits that supply bedroom receptacles be protected by an arc-fault circuit interrupter.

3-65 What is the minimum size copper wire that can be used on a 600-volt system?

A) 14 AWG

B) 12 AWG

C) 10 AWG

D) 8 AWG

Answer: A

NEC Table 310.5 gives the minimum size copper conductor as No. 14 AWG (No. 12 AWG for aluminum) for use on systems from 0 through 2000 volts.

3-66 The maximum overcurrent protection allowed on No. 14 AWG copper THWN wire when used in a raceway for branch circuits is:

A) 30 amperes

B) 20 amperes

C) 15 amperes

D) 10 amperes

Answer: C

NEC Section 240.4(D) specifies 15 amperes.

3-67 The maximum overcurrent protection allowed on No. 12 AWG copper THWN wire when used in a raceway for branch circuits is:

A) 30 amperes

B) 20 amperes

C) 15 amperes

D) 10 amperes

Answer: B

NEC Section 240.4(D) specifies 20 amperes.

3-68 The maximum overcurrent protection allowed on No. 10 AWG copper THWN wire when used in a raceway for branch circuits is:

A) 30 amperes

B) 20 amperes

C) 15 amperes

D) 10 amperes

Answer: A

NEC Section 240.4(D) specifies 30 amperes.

3-69 Which of the following locations prohibits the use Type MI cable?

A) For residential branch circuits

B) Where exposed to excessive moisture

C) Where exposed to destructive corrosive conditions

D) Where exposed to low temperature

Answer: C

Highly corrosive conditions can cause deterioration of the metal jacket. NEC Section 332.12(2).

3-70 Electrical nonmetallic tubing (ENT) may not be used:

A) In hazardous locations

B) For the support of fixtures and other equipment

C) Where subject to ambient temperatures exceeding those for which the tubing is listed

D) All of the above

Answer: D

ENT may not be used in any of the locations listed. Furthermore, ENT may not be used for direct burial in the earth, where voltage exceeds 600 volts, in theaters and similar locations, and other places as described in NEC Section 362.12.

3-71 Splices in an ENT system may be made only at:

A) Connectors

B) Junction boxes

C) The beginning of an ENT system

D) 12 inches from outlet boxes

Answer: B

NEC Section 362.56 specifies that splices and taps may be made only in junction boxes, outlet boxes or conduit bodies. Also see NEC Article 300.15.

3-72 Where practicable, dissimilar metals in contact anywhere in the system shall be avoided to eliminate the possibility of _______.

A) Hysteresis

B) Specialty gravity

C) Galvanic action

D) Resistance

Answer: C

NEC Section 344.14 requires protection of metal conduits from corrosive influences.

3-73 Type MC (metal-clad) cable may not be used in:

A) Residential occupancies

B) Industrial applications

C) Commercial applications

D) Areas where the cable is exposed to corrosive conditions

Answer: D

NEC Section 330.12 does not permit Type MC cable to be used in areas where corrosive conditions exist because such areas will cause deterioration of the cable's jacket.

3-74 Type NMC cable may be used in:

A) Storage battery rooms

B) Dry or moist locations

C) Hazardous (classified) locations

D) Theaters, auditoriums or similar places of assembly

Answer: B

NEC Section 334.10(B) permits Type NMC cable to be installed in the areas described in Answer B; that is, dry or moist locations.

3-75 Power and control tray cable (Type TC) may not be used in:

A) Areas where the cable will be exposed to physical damage

B) Outdoor areas when supported with messenger cable

C) Circuits intended for signal circuits

D) Circuits used for power and lighting

Answer: A

NEC Section 336.12 prohibits the use of Type TC cable in areas where the cable will be exposed to physical damage.

3-76 What is the minimum size IMC conduit allowed for electrical systems for building construction?

A) 2 inches

B) 1 inch

C) 3/4 inch

D) 1/2 inch

Answer: D

NEC Section 342.20 prohibits the use of IMC conduit smaller than 1/2 inch.

3-77 What is the maximum continuous load (in amperes) that can be used on a 240-volt, 20-ampere circuit using two No. 12 AWG conductors with THW insulation when the circuit is supplied by an assembly together with its overcurrent device that is listed for continuous operation at 100 percent of its rating?

A) 12 amperes

B) 15 amperes

C) 16 amperes

D) 20 amperes

Answer: D

A circuit supplied by an assembly together with its overcurrent device that is listed for continuous operation at 100 percent of its rating may carry the full load. NEC Section 210.20(A), Exception. Since a No. 12 conductor is rated at 20 amperes, the maximum current permitted for continuous loads is 20 amperes.

3-78 What is an AFCI?

A) Ampere faulting capacitor insulator

B) Alternate fire collector interrupter

C) Appliance and fixture circuit identification

D) Arc-fault circuit-interrupter

Answer: D

Section 210.12(A) defines an arc-fault circuit-interrupter as a device intended to provide protection from the effects of arc faults by recognizing characteristics unique to arcing and by functioning to de-energize the circuit when an arc fault is detected.

3-79 Where a cable or raceway wiring method is installed through holes bored in wooden floor joists, rafters, or studs and the hole is less than 1¼ inches from the nearest edge of the wood member, what protection must be provided?

A) A steel plate or bushing

B) Duct tape

C) The raceway must be installed in PVC rigid conduit

D) The installation must be removed and installed elsewhere

Answer: A

NEC Section 300.4(A)(1) requires steel plates or bushings for protection.

3-80 Where steel plates are used to protect a cable or wiring method as described in the above question, what is the required minimum thickness of the plate?

A) $^1/_{32}$ inch

B) $^1/_{16}$ inch

C) $^1/_8$ inch

D) $^3/_8$ inch

Answer: B

NEC Section 300.4(A)(1) requires a minimum thickness of $^1/_{16}$ inch.

3-81 Which of the following locations require GFCI protection for all 125-volt, 15- and 20-ampere receptacles?

A) In residential attics

B) Within 7 feet of a residential wet bar

C) A bathroom in a commercial establishment with a wash basin and toilet

D) Garage receptacles that are not readily accessible

Answer: C

NEC Section 210.8(B)(1) requires GFCI protection in bathrooms of commercial, industrial, and all other nondwelling occupancies. All residential bathrooms must also be provided with GFCI as prescribed in NEC Section 210.8(A)(1).

3-82 Which of the following must be installed at or near equipment requiring servicing such as HVAC equipment in attics or crawl spaces?

A) Three duplex receptacles, all with GFCI protection

B) Lighting fixture with pull-chain switch

C) Low-voltage transformer to service control components

D) At least one switch-controlled lighting outlet

Answer: D

NEC Sections 210.70(A) and (C) require at least one switch-controlled lighting outlet so that service personnel can turn on a light prior to entering the work area.

3-83 In the preceding question, where must the switch be located?

A) At the usual point of entry to the area containing the equipment

B) Within 6 feet of the equipment being serviced

C) In another room separated by one lockable door

D) In a metal housing with provisions for locking

Answer: A

NEC Sections 210.70(A) and (C) continue with the requirement of having the switch at the usual point of entry so that service personnel can turn on a light prior to entering the work area.

3-84 What is the maximum size rigid metal conduit allowed for electrical construction?

A) 2 inches

B) 4 inches

C) 6 inches

D) 8 inches

Answer: C

NEC Section 344.20(B) prohibits the use of conduit larger than 6 inches.

3-85 The *NEC* defines "unfinished basement" as portions or areas of the basement:

A) Intended as habitable rooms

B) Limited to storage or work areas

C) Limited to recreation only

D) Used for storing vehicles

Answer: B

NEC Section 210.8(A)(5) defines "unfinished basement" as: portions or areas of the basement not intended as habitable rooms and limited to storage areas, work areas, and the like.

3-86 Which of the following locations requires ground-fault circuit-interrupter protection for personnel on all 125 volt, single-phase, 15- and 20-ampere receptacles?

A) Installed less than 4 feet, 6 inches above the floor inside of the building

B) Installed in an attic stairway

C) Installed outdoors

D) Installed within 8 feet of kitchen sink

Answer: C

NEC Section 210.8(A)(3) requires all residential outdoor receptacles to be provided with GFCI protection.

3-87 Appliance outlets installed for a specific appliance must be installed within how many feet of the intended location of the appliance?

A) 6 feet

B) 8 feet

C) 10 feet

D) 12 feet

Answer: A

NEC Section 210.50(C) requires that the outlet be installed within 6 feet of the intended location of the appliance.

Chapter 4

Electric Services

Electric services can range in size from a small 120-volt, single-phase, 15-ampere service — the minimum allowed by Section 230.79(A) of the *National Electrical Code®* — to huge industrial installations involving substations dealing with thousands of volts and amperes. Regardless of the size, all electric services are provided for the same purpose: for delivering electrical energy from the supply system to the wiring system on the premises served. Consequently, all establishments containing equipment that utilizes electricity require an electric service.

Figure 4-1 on the next page shows the basic sections of a typical electric service. In this illustration, note that the high-voltage lines terminate on a power pole near the building that is being served. A transformer is mounted on the pole to reduce the voltage to a usable level (120/240 volts in this case). The remaining parts of the service consist of a service drop, a service-entrance, service-entrance conductors and service-entrance equipment. A description of each follows:

- *Service drop:* The overhead conductors, through which electrical service is supplied, between the last power company pole and the point of their connection to the service facilities located at the building or other support used for the purpose.
- *Service entrance:* All components between the point of termination of the overhead service drop or underground service lateral and the building's main disconnecting device, except for metering equipment.
- *Service conductors:* The conductors between the point of termination of the overhead service drop or underground service lateral and the main disconnecting device.
- *Service equipment:* Provides overcurrent protection to the feeder and service conductors, a means of disconnecting the feeders from energized service conductors, and a means of measuring the energy used by the use of metering equipment.

When the service conductors to the building are routed underground, these conductors are known as the service lateral, defined as follows:

- *Service lateral:* The underground conductors through which service is supplied between the power company's distribution facilities and the first point of their connection to the building or area service facilities.

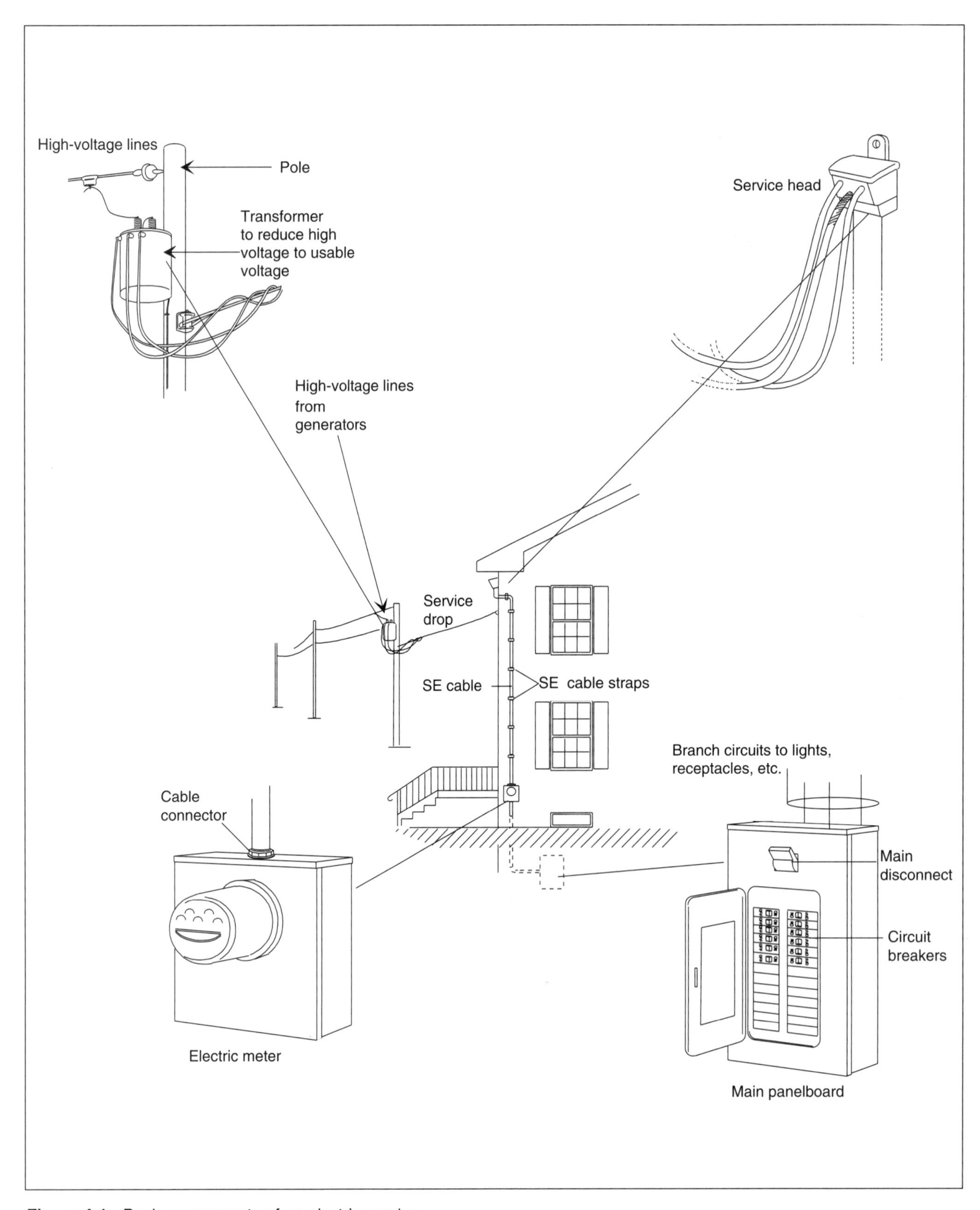

Figure 4-1: Basic components of an electric service

4-1 The overhead conductors, through which electrical service is supplied, between the last power company pole and the building or other support used for the purpose are called:

A) Service entrance
B) Service-entrance conductors
C) Service drop
D) Service-entrance equipment

Answer: C

NEC Article 100 — Definitions. See Figure 4-1.

4-2 The conductors and equipment for delivering electric energy from the power company to the wiring system of the premises served is called a:

A) Service
B) Service-entrance conductors
C) Service drop
D) Service-entrance equipment

Answer: A

NEC Article 100 — Definitions. See Figure 4-1.

4-3 The conductors between the point of termination of the overhead service drop or underground service lateral and the main disconnecting device in a building are known as:

A) Service
B) Service conductors
C) Service drop
D) Service-entrance equipment

Answer: B

NEC Article 100 — Definitions. See Figure 4-1.

4-4 The necessary equipment connected to the load end of the service conductors to a building or other structure and intended to constitute the main control and cutoff of the supply is known as:

A) Service entrance
B) Service-entrance conductors
C) Service drop
D) Service equipment

Answer: D

NEC Article 100 — Definitions. See Figure 4-1.

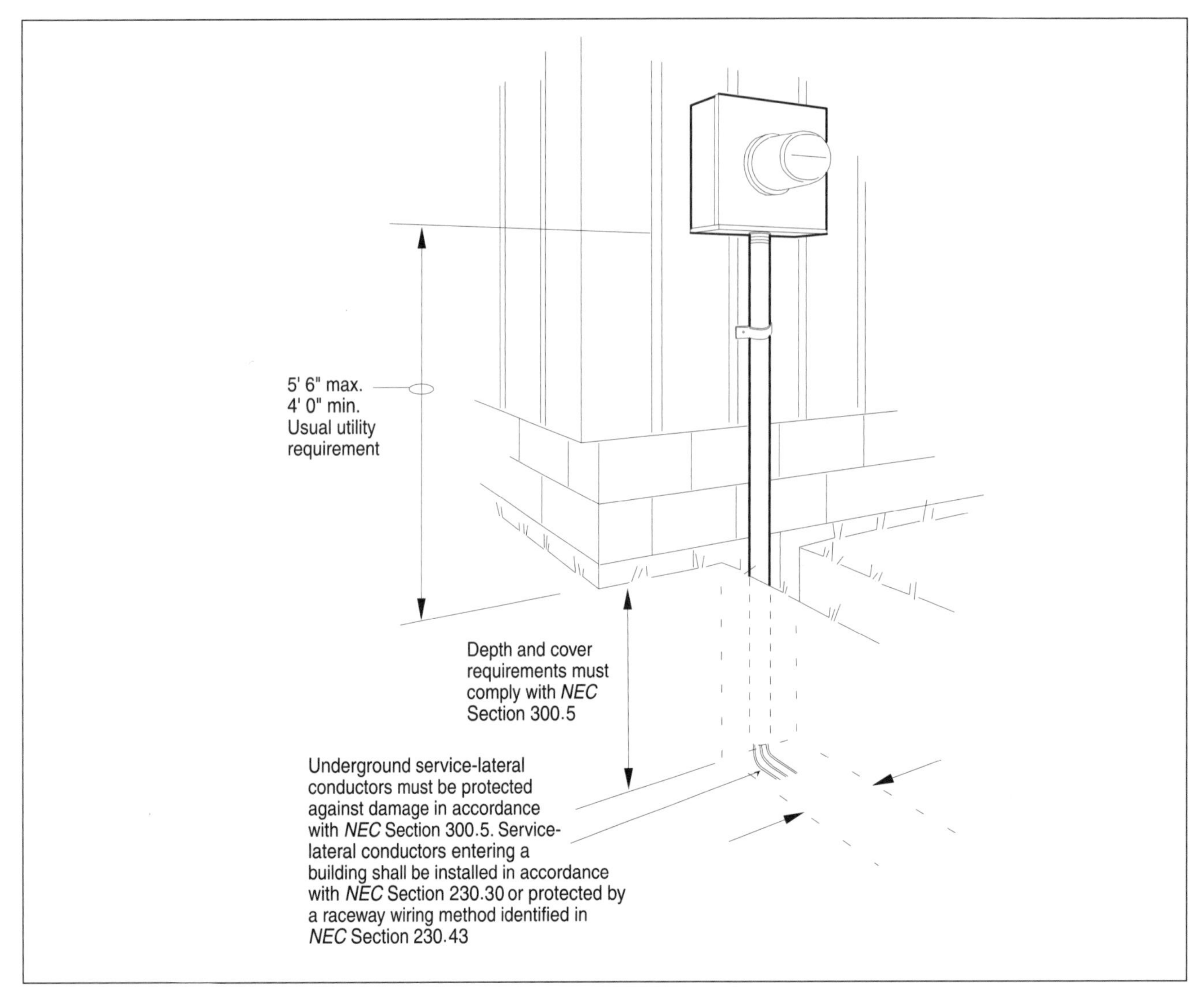

Figure 4-2: Underground service lateral

4-5 The underground conductors through which service is supplied between the power company's distribution facilities and the first point of their connection to a building or area service facilities located at the building or other support used for the purpose are known as:

A) Service clearance

B) Service lateral

C) Parallel connection

D) Weatherhead

Answer: B

Underground electric services frequently emanate from a pad-mounted transformer, but the service conductors can be run down a power pole in conduit to a point underground. From this point, the conductors run underground to the building. NEC Article 100 — Definitions. See Figure 4-2.

4-6 In a balanced three-phase, 4-wire system, the current in the neutral conductor:

A) Is the same as in the other conductors

B) Is lower than in the other conductors

C) Is higher than in the other conductors

D) Will not flow when all loads are exactly balanced

Answer: D

The grounded or neutral conductor carries only the unbalanced load. If, say, a single-phase, 120/240-volt service carries 100 amperes on phase A, and 80 amperes on phase B, the neutral would carry only 20 amperes — the difference between the current in phase A and that in phase B.

4-7 On a 240-volt, single-phase, 3-wire system, the neutral conductor will:

A) Never carry current even if the other conductors carry current

B) Never carry current larger than the difference between the current in the two ungrounded ("hot") conductors

C) Carry current equal to the current between phases

D) Carry current equal to one-half the current between phases

Answer: B

See explanation to question No. 4-6 above.

4-8 The unit used to measure current is the:

A) Ohm

B) Ampere

C) Watt

D) Coulomb

Answer: B

Current is the rate at which electricity flows through a conductor. The unit of measurement is the "ampere." The relationship between ohms, watts, amperes, and voltage is expressed in Ohm's Law. See Figure 4-3.

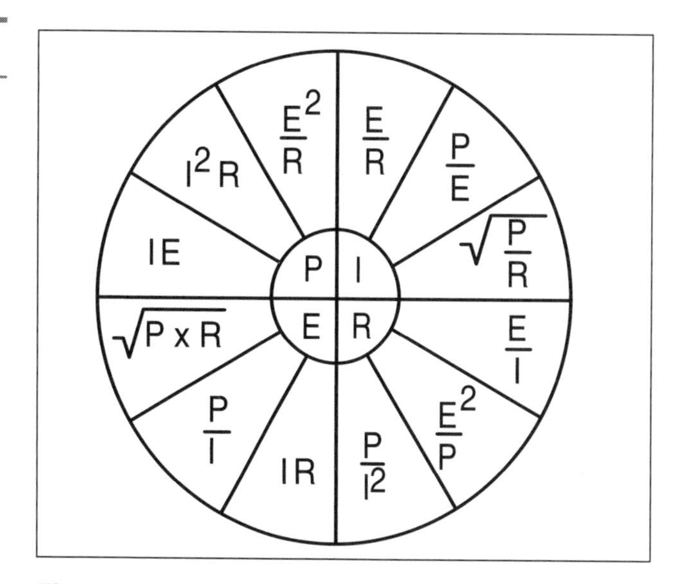

Figure 4-3: Summary of Ohm's law

4-9 The minimum allowable current rating of 3-wire service-entrance conductors for a one-family dwelling is:

A) 60 amperes

B) 100 amperes

C) 125 amperes

D) 150 amperes

Answer: B

NEC Section 230.42 states that the ungrounded conductors for specific installations not be smaller than the disconnecting means. NEC 230.79(C) requires the service disconnecting means for a one-family dwelling not be less than 100 amperes.

4-10 The minimum allowable current rating of 3-wire service-entrance conductors for a one-family dwelling with an initial net computed load of 10 kVA or more is:

A) 60 amperes

B) 100 amperes

C) 125 amperes

D) 150 amperes

Answer: B

Regardless of the computed load or the number of circuits, a one-family dwelling shall have a minimum service size of 100 amperes. NEC 230.79(C).

4-11 What is the smallest copper wire size allowed for service-entrance conductors?

A) No. 10 AWG

B) No. 8 AWG

C) No. 12 AWG

D) No. 4 AWG

Answer: B

Service-entrance conductors shall not be smaller than No. 8 copper or No. 6 aluminum. NEC Section 230.31(B).

4-12 The smallest grounded or neutral conductor for an electric service using 1100 kcmil copper conductors is:

A) 2/0 copper
B) 3/0 copper
C) 1/0 copper
D) No. 4 AWG copper

Answer: A

For service-entrance conductors, the grounded conductor shall not be smaller than indicated in NEC Table 250.66.

4-13 What is the smallest aluminum or copper-clad wire size allowed for service-entrance conductors?

A) No. 10 AWG
B) No. 12 AWG
C) No. 6 AWG
D) No. 4 AWG

Answer: C

The smallest size wire that can be used as service-entrance conductor is No. 8 copper or No. 6 aluminum. NEC Section 230.31(B).

4-14 The name given to a system or circuit conductor that is intentionally grounded is:

A) Grounding conductor
B) Bonding conductor
C) High-leg conductor
D) Grounded or neutral conductor

Answer: D

An intentionally grounded conductor is a grounded conductor as explained in NEC Article 100 — Definitions.

4-15 To measure the area of a dwelling to determine the lighting load, the following dimensions are used:

A) The floor area computed from the inside dimensions

B) The cubic feet of each room

C) The floor area computed from the outside dimensions

D) The area six inches from the inside walls

Answer: C

The floor area for each floor must be computed from the outside dimensions of the building, apartment, or other area involved. NEC Section 220.3(A). See Figure 4-4.

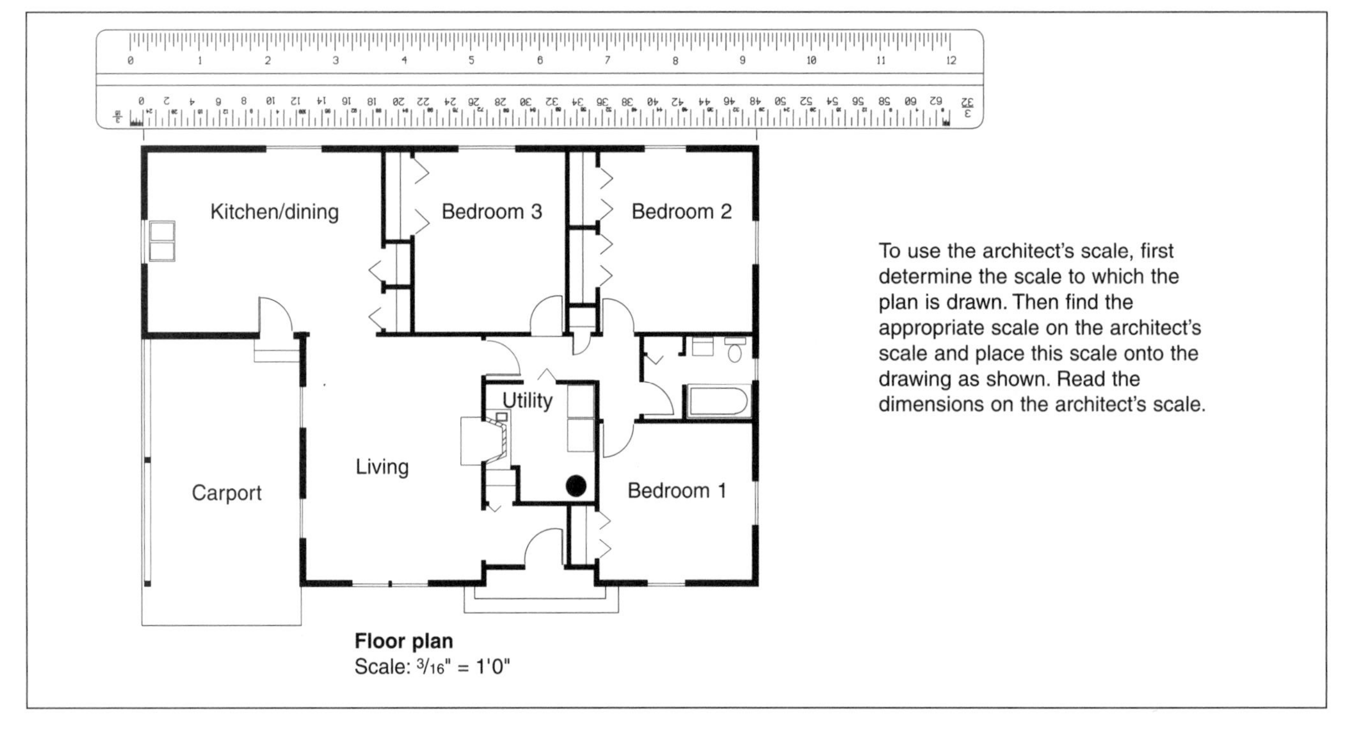

Figure 4-4: Measurements are taken from the outside dimensions of the building

4-16 What is the smallest copper wire size allowed for service-entrance conductors supplying loads consisting of limited loads of a single branch circuit?

A) No. 10 AWG

B) No. 12 AWG

C) No. 4 AWG

D) No. 6 AWG

Answer: B

For limited loads of a single branch circuit, No. 12 copper may be used, but in no case smaller than the branch-circuit conductors. NEC Section 230.31(B), Exception.

4-17 The allowable number of service disconnects can consist of up to:

A) Three
B) Four
C) Five
D) Six

Answer: D

The service disconnection means shall consist of not more than six switches or six circuit breakers. NEC Section 230.71(A).

4-18 Service conductors that pass over a flat roof must have what clearance from the highest point of roof over which they pass?

A) 4 feet
B) 6 feet
C) 8 feet
D) 10 feet

Answer: C

NEC Section 230.24(A). *See Figure 4-5 on the next page.*

4-19 Where the voltage between conductors does not exceed 300 and the roof has a slope of not less than 4 inches in 12 inches, the minimum clearance of service conductors is:

A) 3 feet
B) 4 feet
C) 5 feet
D) 6 feet

Answer: A

NEC Section 230.24(A), Exception No. 2. *See Figure 4-5 on the next page.*

4-20 Where the voltage between conductors does not exceed 300, and not more than 6 feet of service conductors pass over not more than 4 feet of the overhang portion of the roof, and the conductors terminate in a through-the-roof raceway or other approved support, the clearance of service conductors above the roof is reduced to:

A) 6 inches
B) 12 inches
C) 18 inches
D) 24 inches

Answer: C

NEC Section 230.24(A), Exception No. 3. *See Figure 4-5 on the next page.*

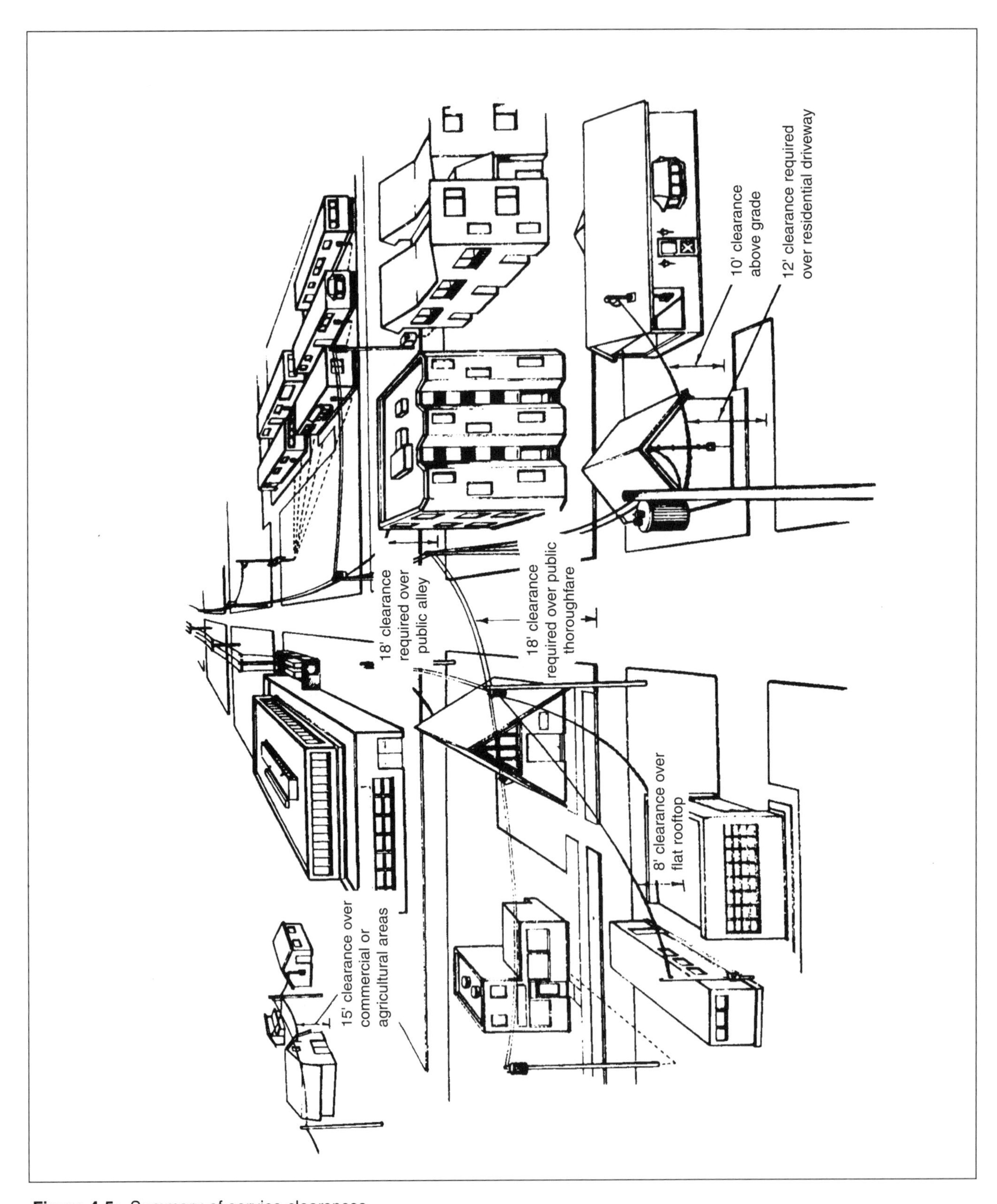

Figure 4-5: Summary of service clearances

4-21 Service-drop conductors, when not in excess of 150 volts to ground, that pass over residential sidewalks, finished grade (the ground), or over any platform or any projection from which the conductors may be reached, must have a minimum clearance of:

A) 8 feet
B) 10 feet
C) 12 feet
D) 16 feet

Answer: B

NEC Section 230.24(B). See Figure 4-5.

4-22 Service-drop conductors, when not in excess of 300 volts, that pass over residential driveways and commercial areas such as parking lots and drive-in establishments not subject to truck traffic must have a clearance of at least:

A) 8 feet
B) 10 feet
C) 12 feet
D) 16 feet

Answer: C

NEC Section 230.24(B). See Figure 4-5.

4-23 Service-drop conductors, when not in excess of 600 volts, that pass over public streets, alleys, roads, parking lots, agricultural, or other areas subject to truck traffic must have a minimum clearance of:

A) 10 feet
B) 12 feet
C) 15 feet
D) 18 feet

Answer: D

NEC Section 230.24(B). See Figure 4-5.

4-24 Service-drop conductors, when not in excess of 600 volts, that pass over residential property, driveways, and those commercial areas not subject to truck traffic must have a minimum clearance of:

A) 10 feet

B) 12 feet

C) 18 feet

D) 15 feet

Answer: D

NEC Section 230.24(B). See Figure 4-5.

4-25 Multiconductor cables used for service drops must be attached to buildings or other structures by:

A) Tie wires

B) Fittings approved for the purpose

C) 16d nails or larger

D) Tapcons

Answer: B

An approved fitting, such as an insulator, is required for all such installations. NEC Section 230.27.

4-26 Service heads for overhead service-entrance conductors must be:

A) Explosionproof

B) Raintight

C) Waterproof

D) Dustproof

Answer: B

Service heads must be constructed and protected to prevent rain from interfering with the successful operation of the service-entrance conductors. NEC Section 230.54(A).

4-27 Service-entrance cable must be supported within what distance from the service head?

A) 12 inches

B) 18 inches

C) 24 inches

D) 36 inches

Answer: A

An SE cable strap, or other approved fitting, must be installed within 12 inches of the service head. NEC Section 230.51(A).

4-28 Service-entrance cable must also be supported at intervals of:

A) 2 feet

B) 2½ feet

C) 3 feet

D) 4½ feet

Answer: B

Type SE cable must be supported at least every 2½ feet (30 inches) to prevent sag or collapse of the cable. NEC Section 230.51(A).

4-29 What clearance must service conductors have from windows, doors, porches, fire escapes, or similar locations?

A) 1 foot

B) 2 feet

C) 3 feet

D) 4 feet

Answer: C

Service conductors installed as open conductors or multiconductor cable without an overall outer jacket shall have a clearance of not less than 3 feet from windows, doors, porches, fire escapes, or similar locations. Conductors run above the top level of a window, however, shall be permitted to be less than 3 feet. NEC Section 230.9.

4-30 When installing a service head, the conductors of different potential should be:

A) Brought out through separately bushed openings

B) Brought out through the same bushed openings

C) Brought out through the same unbushed openings

D) Brought out through separately unbushed openings

Answer: A

Service heads shall have conductors of opposite potential brought out through separately bushed openings. *NEC Section 230.54(E). See detail of service head in Figure 4-1.*

4-31 Metal enclosures for service conductors and equipment must:

A) Never be grounded
B) Be grounded
C) Be coated with non-conductive material
D) Be constructed of at least 38 gauge steel

Answer: B

All metal enclosures or housings for service conductors, such as meter bases, auxiliary gutters, panelboards, safety switches, etc. must be grounded with the system's equipment ground. NEC Section 250.80.

4-32 What size service is required for a 120/240-volt, single-phase load with a demand load of 36 kVA?

A) 60 amperes
B) 100 amperes
C) 150 amperes
D) 200 amperes

Answer: C

Divide the voltage between phases into the total volt-amperes or watts. Thus, 36,000/240 = 150 amperes.

4-33 What size copper THW conductors are required for the service load in Question 4-32?

A) 1/0
B) 2/0
C) 3/0
D) 4/0

Answer: A

NEC Table 310.16 shows that 1/0 THW wire has a current-carrying capacity of 150 amperes when the ambient temperature is 75°C or below.

4-34 What size rigid steel conduit is required to contain the three conductors in Question 4-33?

A) 1 1/2 inch
B) 2 inch
C) 2 1/2 inch
D) 3 inch

Answer: A

Table C.8 in NEC Annex C shows that 1 1/2 inch conduit can contain three No. 1/0 THW conductors without exceeding the 40% fill requirements.

4-35 What percentage of fill is allowed in a conduit with three conductors?

A) 30%
B) 40%
C) 50%
D) 60%

Answer: B

When three or more conductors with any type insulation, other than lead-covered, are installed in a raceway, the conductors may not exceed 40% of the raceway's cross-sectional area. NEC Chapter 9, Table 1.

4-36 If THHN copper conductors are used for the service-entrance in Question 4-32, what size is required?

A) No. 4 AWG
B) No. 3 AWG
C) No. 2 AWG
D) No. 1 AWG

Answer: D

Due to the insulation on THHN conductors, they are able to stand a higher temperature and therefore have a greater current-carrying capacity. NEC Table 310.16 or Table 310.15(B)(6) when applicable.

4-37 What size electrical metallic tubing (EMT) is required to contain the service conductors in Question 4-36?

A) 1 inch
B) 1 1/4 inch
C) 1 1/2 inch
D) 2 inch

Answer: B

1 1/4 inch EMT is sufficient to handle three No. 1 THHN conductors without exceeding the required 40% fill. NEC Table C.8, Annex C.

4-38 When using $2^1/_2$-inch rigid steel conduit for a through-the-roof 120/240 volt service mast, what is the minimum distance the conduit can protrude above the roof?

A) 18 inches

B) 24 inches

C) 36 inches

D) 48 inches

Answer: A

NEC Section 230.24(A) Exception No. 3 allows a reduction in clearance above only the overhanging portion of the roof to not less than 18 inches.

4-39 What is the minimum size copper grounding electrode conductor allowed on an electric service utilizing No. 3/0 copper current-carrying conductors?

A) No. 8 AWG

B) No. 6 AWG

C) No. 4 AWG

D) No. 2 AWG

Answer: C

NEC Table 250.66 specifies a No. 4 copper or No. 2 aluminum conductor as the minimum size grounding electrode conductor.

4-40 The main service disconnecting means should be located:

A) At or near the point where the service-entrance conductors enter the building

B) At least 20 feet from the point where the service-entrance conductors enter the building

C) At least 30 feet from the point where the service-entrance conductors enter the building

D) At least 50 feet from the point where the service-entrance conductors enter the building

Answer: A

The service disconnecting means must be installed at a readily accessible location either outside of a building or structure, or inside, nearest the point of entrance of the service conductors. NEC Section 230.70(A).

4-41 The *NEC* requires that all circuits over 150 volts to ground containing fuses have a disconnecting means located:

A) Outside

B) On supply side of all fuses

C) On load side of all fuses

D) Where convenient

Answer: B

Disconnecting means must be provided on the supply side of all fuses in circuits of over 150 volts to ground. NEC Section 240.40.

4-42 The general lighting load for residential services is calculated at:

A) 1 watt per square foot

B) 2 watts per square foot

C) 3 watts per square foot

D) 4 watts per square foot

Answer: C

NEC Table 220.12 shows that the general lighting load for dwelling units is calculated at 3 volt-amperes per square foot.

4-43 Residential small appliance circuits are calculated at:

A) 1500 watts each

B) 2000 watts each

C) 2500 watts each

D) 3000 watts each

Answer: A

In addition to the number of branch circuits determined by the square-foot method (based on 3 volt-amperes per square foot), two or more 20-ampere small appliance branch circuits must be provided for all receptacle outlets in the kitchen, pantry, breakfast room, dining room, or similar area of a dwelling unit. NEC Section 220.52(A).

4-44 All metal enclosures for service conductors and equipment must be:

A) Watertight

B) Painted

C) Locked

D) Grounded

Answer: D

NEC Section 250.80 states that metal enclosures for service conductors and equipment must be grounded.

4-45 Mobile home service equipment shall not be rated less than:

A) 50 amperes

B) 60 amperes

C) 100 amperes

D) 200 amperes

Answer: C

Mobile home service equipment shall be rated at not less than 100 amperes, and provision must be made for connecting a mobile home feeder assembly by a permanent wiring method. NEC Section 550.32(C).

4-46 What is the ampacity of the conductors required for a 120/208-volt, three-phase, 4-wire commercial service with a load of 72 kVA?

A) 150 amperes

B) 200 amperes

C) 250 amperes

D) 300 amperes

Answer: B

$$Amperes = \frac{VA}{Volt \times \sqrt{3}} = \frac{72{,}000}{208 \times \sqrt{3}} = 200 \text{ amperes}$$

4-47 What size aluminum THW conductors are required for the service in Question 4-46?

A) 1/0 kcmil

B) 3/0 kcmil

C) 4/0 kcmil

D) 250 kcmil

Answer: D

NEC Table 310.16 shows that 250 kcmil THW aluminum conductors are rated at 205 amperes.

4-48 What size rigid steel conduit is required for the service conductors in Question 4-47?

A) 2 inch

B) 2½ inch

C) 3 inch

D) 4 inch

Answer: B

NEC Table C.8 in Annex C specifies a 2½-inch conduit for four 250 kcmil THW conductors.

4-49 What is the maximum distance between supports for the conduit in Question 4-48 if the conduit is a straight run?

A) 2 feet

B) 4 feet

C) 8 feet

D) 16 feet

Answer: D

NEC Table 344.30(B)(2) specifies a maximum distance of 16 feet between $2^1/2$ inch rigid conduit supports. See Figure 4-6.

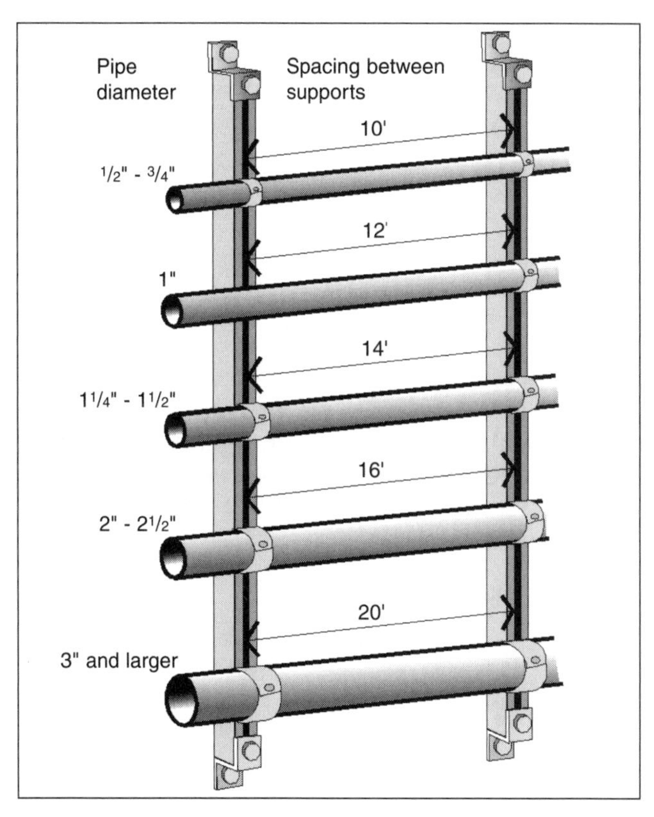

Figure 4-6: Spacing of supports for rigid steel conduit

4-50 What is the maximum distance for a required support for rigid steel conduit from the service head?

A) 3 feet

B) 2 feet

C) 4 feet

D) 5 feet

Answer: A

NEC Section 344.30(A) states that rigid metal conduit shall be firmly fastened within 3 feet of each outlet box, junction box, cabinet, or fitting. Since a service head is classified as a fitting, the maximum distance is 3 feet.

4-51 What is the required ampacity of conductors serving a 120/240-volt single-phase residential electric service with a total load of 30 kVA?

A) 100 amperes

B) 125 amperes

C) 150 amperes

D) 200 amperes

Answer: B

30,000/240 = 125

4-52 What size aluminum SE cable is required for the service in Question 4-51?

A) No. 2 AWG
B) No. 1 AWG
C) No. 1/0 AWG
D) No. 2/0 AWG

Answer: C

NEC Table 310.16.

4-53 What is the maximum distance that an SE cable strap can be from either the service head or the meter base?

A) 10 inches
B) 12 inches
C) 18 inches
D) 24 inches

Answer: B

Cable shall be secured in place at intervals not exceeding 12 inches from every cabinet, box, or fitting. NEC Section 230.51(A).

4-54 What is the maximum distance allowed between cable straps when installing Type SE cable when used as service-entrance conductors?

A) 1 foot
B) 2 1/2 feet
C) 3 feet
D) 4 feet

Answer: B

Cable shall be secured in place at intervals not exceeding 2 1/2 feet. NEC Section 230.51(A).

4-55 When a single-family dwelling has an initial load of 10 kVA or above, the minimum service rating is:

A) 60 amperes
B) 100 amperes
C) 125 amperes
D) 150 amperes

Answer: B

For a one-family dwelling, the service disconnecting means shall have a rating of not less than 100 amperes. NEC Section 230.79(C).

4-56 If the phase-to-neutral voltage in a three-phase, 4-wire Y-connected service is 240 volts, the phase-to-phase voltage will be approximately:

A) 460 volts
B) 480 volts
C) 415 volts
D) 425 volts

Answer: C

Phase-to-phase voltage = phase-to-neutral voltage $\times \sqrt{3} = 240 \times \sqrt{3} = 240 \times 1.73 = 415$ *volts*

4-57 The service grounding conductor is sized by the rating of:

A) The overcurrent protective device
B) The service-entrance conductors
C) The supply transformer
D) The load to be served

Answer: B

NEC Section 250.66 states that the size of the grounding electrode conductor of a grounded or ungrounded ac system shall not be less than the sizes given in NEC Table 250.66. The sizes of grounding electrode conductors are based on the size of the largest service conductor. Both copper and aluminum service conductors are listed in this table.

4-58 A 400-ampere electric service is normally metered with a combination of a watt-hour meter and:

A) Current transformers
B) VAR meters
C) Capacitors
D) Ammeter

Answer: A

A watt-hour meter constructed to directly measure currents over 200 amperes is too expensive to be practical. Therefore, current transformers are used to reduce the ratio of the current to a practical level.

4-59 A 120/240-volt residential electric service requires No. 2 AWG copper for the ungrounded conductors. What is the minimum size neutral conductor allowed?

A) No. 10 AWG

B) No. 8 AWG

C) No. 6 AWG

D) No. 4 AWG

Answer: B

The neutral conductor must not be smaller than the required grounding electrode conductor specified in NEC Table 250.66. NEC Section 250.24(B)(1).

4-60 When two to six fused switches or circuit breakers are used as the "main" in an electric service, they must be:

A) Grouped and marked to indicate the load served

B) Grouped but left unmarked

C) Marked

D) Grouped

Answer: A

NEC Section 230.72(A) requires that two to six disconnects as permitted in NEC Section 230.71 must be grouped and each disconnect must be marked to indicate the load served.

4-61 In a multiple-occupancy building, each occupant shall:

A) Not have access to the occupant's service disconnecting means

B) Have access to the occupant's service disconnecting means

C) Have access to all occupants' service disconnecting means

D) Have complete access to all electrical equipment

Answer: B

NEC Section 230.72(C) specifies that in a multiple-occupancy building, each occupant shall have access to the occupant's service disconnecting means.

4-62 Although enclosed, a service disconnecting means must still be:

A) Locked securely

B) Always kept unlocked

C) Locked only while operable

D) Manually or power operable

Answer: D

NEC Section 230.76 specifies that an enclosed service disconnecting means shall consist of either a manually operated switch or circuit breaker equipped with a handle or a power-operated switch or circuit breaker which can be opened by hand in the event of a power failure.

4-63 The service disconnecting means for a two-circuit installation shall not be less than:

A) 15 amperes

B) 20 amperes

C) 30 amperes

D) 60 amperes

Answer: C

NEC Section 230.79(B) states that for installations consisting of not more than two 2-wire branch circuits, the service disconnecting means shall have a rating of not less than 30 amperes.

4-64 A means must be provided to disconnect the grounded conductor from the premises wiring. One approved method is:

A) A terminal with pressure connectors

B) A ground clamp on a grounding electrode

C) A wire trough

D) A cablebus

Answer: A

Where the service disconnecting means does not disconnect the grounded conductor from the premises wiring, other means must be provided for this purpose in the service equipment. A terminal or bus to which all grounded conductors can be attached by means of pressure connectors is permitted for this purpose. NEC Section 230.75.

4-65 Service conductors shall *not* be connected to the service disconnecting means with:

A) Pressure connectors

B) Clamps

C) Solder only

D) Approved conductor terminators

Answer: C

Service conductors shall be connected to the service disconnecting means by pressure connectors, clamps, or other approved means. Connections that depend upon solder shall not be used. NEC Section 230.81.

4-66 What is the minimum allowed rating for the service disconnect of a single circuit installation?

A) 15 amps

B) 20 amps

C) 30 amps

D) 10 amps

Answer: A

NEC Section 230.79(A) stipulates that 15 amps is the minimum allowed rating for the service disconnect in this situation.

4-67 Immediately before service conductors enter the bushed holes in a service head, what should be provided for the conductors?

A) Drip loops

B) Solderless connectors

C) Weatherproof connectors

D) Soldered connectors

Answer: A

Drip loops in service conductors help to prevent moisture from entering the service head. NEC Section 230.54(F). See Figure 4-7.

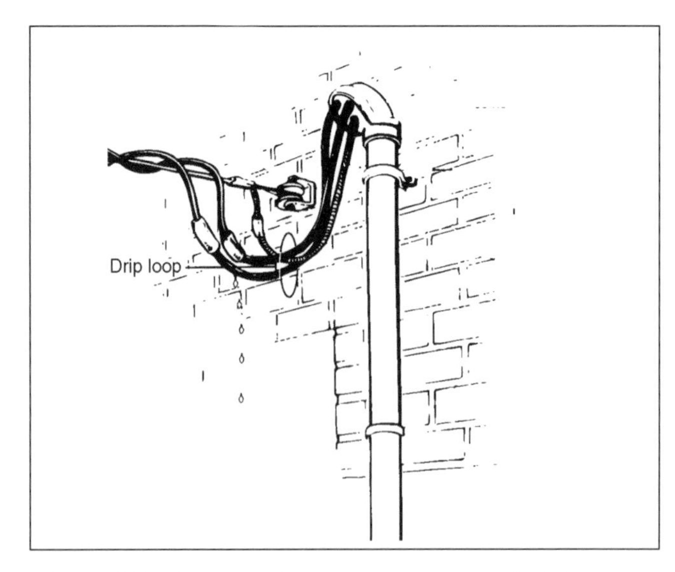

Figure 4-7: Drip loops help keep moisture out of the service head

4-68 Which of the following arrangements is an approved grounding electrode system?

A) A single underground water pipe

B) An indoor metal water pipe fed by PVC pipe from a deep well

C) A single metal underground water pipe used in conjunction with a driven ground rod

D) A single plastic underground water pipe

Answer: C

A single metal water pipe used in conjunction with a driven ground rod is an approved grounding electrode system. Any meters or unions in the water pipe, however, must be bypassed with bonding jumpers in case the pipe is disassembled or becomes loose. NEC Section 250.52(A)(1) and 250.53(D)(2).

4-69 A single made electrode with a resistance to ground of more than 25 ohms is approved as a grounding electrode system only if:

A) The ground rod is driven into the ground 8 feet or more

B) It is augmented by one or more additional electrodes

C) The made electrode is constructed of aluminum

D) The made electrode is constructed of copper

Answer: B

A made electrode, such as a driven ground rod with a resistance to ground of over 25 ohms, must be supplemented by an additional electrode. A metal underground water pipe shall not meet this requirement. NEC Section 250.56.

4-70 If an interior metal water pipe is used as a grounding electrode, what is the maximum distance from the pipe's point of entrance into the building that the grounding conductor must be attached?

A) Within 5 feet

B) Within 10 feet

C) Within 15 feet

D) Within 20 feet

Answer: A

NEC Section 250.52(A)(1) restricts the maximum distance to 5 feet from the point of entrance to the building. *See Figure 4-8 on the next page.*

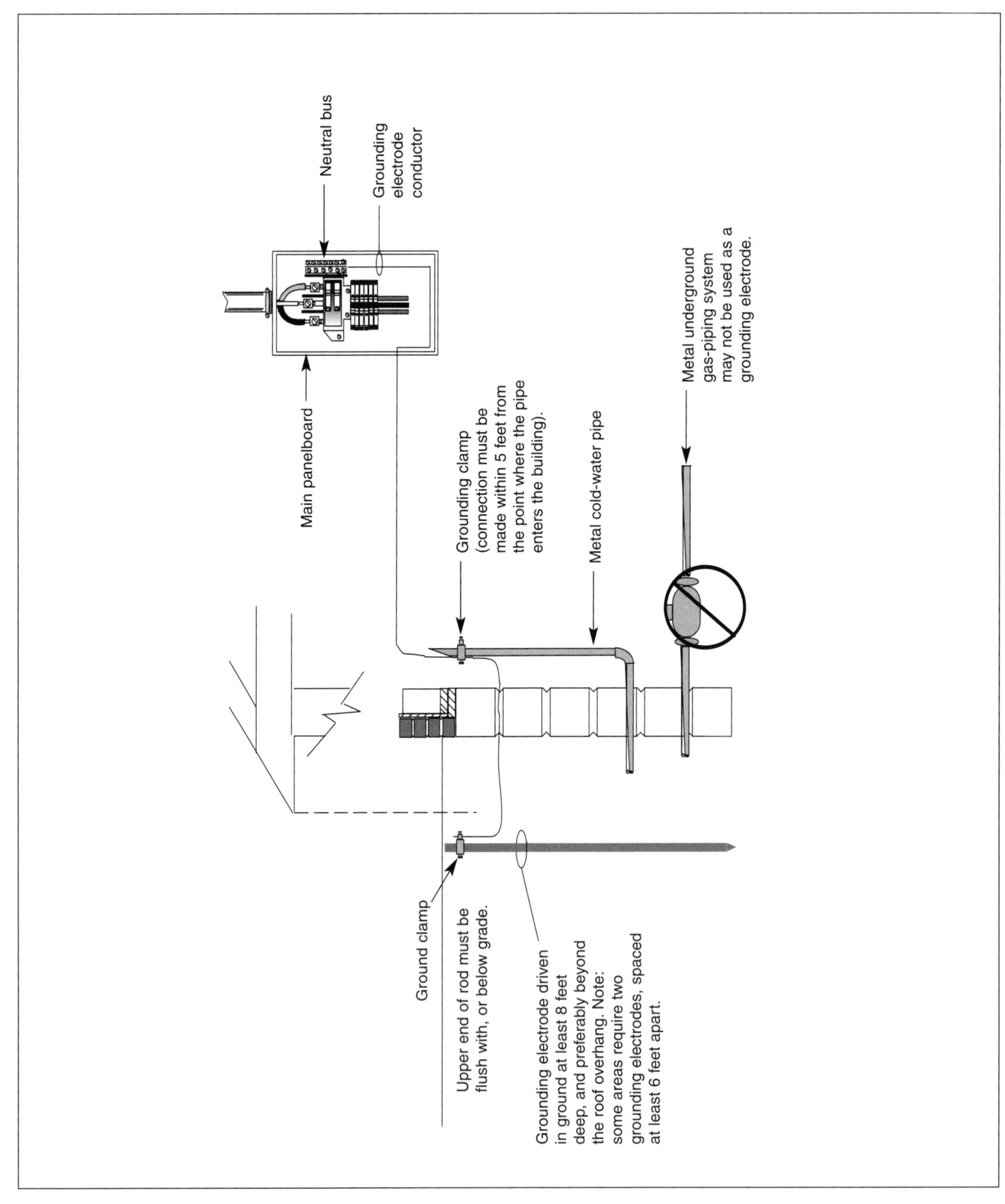

Figure 4-8: Summary of *NEC* grounding requirements

4-71 What is the minimum size neutral conductor allowed on a high-impedance grounded neutral system, carrying a load of 30 amperes, where the conditions of maintenance and supervision assure that only qualified persons will service the installation?

A) No. 10 AWG copper

B) No. 8 AWG copper

C) No. 6 AWG copper

D) No. 4 AWG copper

Answer: B

NEC Section 250.36(B) states that the neutral conductor must have an ampacity of no less than the maximum current rating of the grounding impedance . . . and in no case be smaller than No. 8 (AWG) copper or No. 6 (AWG) aluminum or copper-clad aluminum.

4-72 How must wire terminals be arranged and located in panelboards?

A) So the installer must reach across ungrounded lines

B) So they are not readily accessible

C) So they may not be tampered with or changed

D) So it will not be necessary for service personnel to reach across or beyond an uninsulated ungrounded line bus to make connections

Answer: D

NEC Section 408.3(D) restricts the location of wire terminals to areas where service personnel do not have to reach across or beyond an uninsulated ungrounded line bus to make connections.

4-73 Which of the following may *not* be used as a grounding electrode?

A) Metallic cold-water pipe

B) Driven ground rod

C) Underground metallic gas pipe

D) A grounding ring consisting of No. 2 AWG bare copper

Answer: C

NEC Section 250.52(B) prohibits the use of metallic underground gas piping as a grounding electrode under any circumstances or conditions.

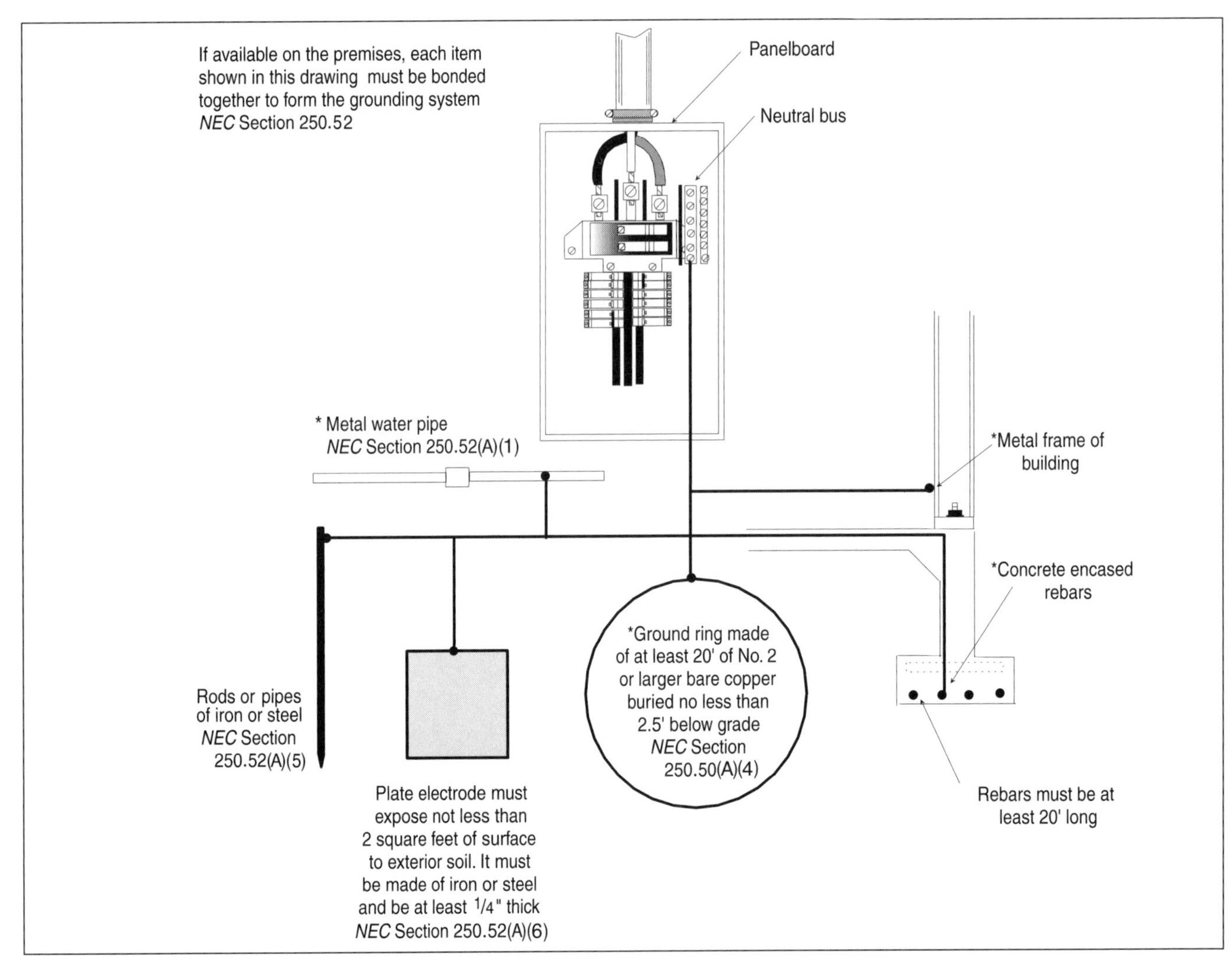

Figure 4-9: Summary of *NEC* grounding requirements

4-74 If the following grounding electrodes are available on the premises, how many of them must be bonded together: metal underground water pipe, metal frame of the building, concrete-encased electrode, ground ring, ground plate, and a ground rod?

A) Only one may be used

B) Two must be used

C) At least three must be bonded together

D) All available electrodes must be bonded together

Answer: D

NEC Section 250.52 requires that all of them be bonded together to form the grounding electrode system. See Figure 4-9.

Chapter 5

Distribution Equipment

Service-entrance equipment is usually grouped at one centralized location. Feeders run to various locations to feed heavy-loaded electrical equipment and subpanels, which are located in a building to keep the length of the branch-circuit runs at a practical minimum for operating efficiency and to cut down on cost.

The main service-disconnecting means will sometimes be made up on the job by assembling individually enclosed fused switches on a length of metal auxiliary gutter. The various components are connected by means of short conduit nipples, in which the insulated conductors are installed. In other cases, the main disconnect and feeder overcurrent devices are enclosed in factory-assembled panelboard; the entire assembly is commonly called a main distribution panel.

Section 230.70 of the *NEC* requires a means to disconnect all conductors in a building or other structure from the service-entrance conductors. This disconnecting means must be installed at a readily accessible location either outside a building or structure, or inside nearest the point of entrance of the service conductors and working space must be provided around the disconnecting means. The *NEC* also requires that all single feeders and branch circuits be provided with a means of individual disconnection from the source of supply.

Overcurrent protection — either fuses or circuit breakers — is required both at the main source and for all individual feeders and branch circuits in order to protect the electrical installation against ground faults and overloads.

Chapter 4 of this book deals with electric services, while Chapter 6 covers questions concerning overcurrent protection. This chapter covers the panelboards, load centers, and similar distribution equipment used at the point the service enters the building and at points throughout the building.

5-1 A lighting and appliance panelboard is a panelboard in which:

A) Ungrounded connections are required to be bolted on

B) Three or more GFCIs are installed

C) 10% of its overcurrent devices are for lighting or appliances

D) Circuit breakers are utilized as switching devices

NEC CHANGE

Answer: C

A lighting and appliance branch-circuit panelboard is one having more than 10% of its overcurrent devices rated 30 amperes or less, for which branch-circuit grounded-conductor connections are provided. NEC Section 408.34.

5-2 All metal enclosures for service conductors and equipment must be:

A) Grounded

B) Ungrounded

C) Made from 10 gauge steel or above

D) PVC (plastic)

Answer: A

Metal enclosures such as metal conduit, pull boxes, panelboard housings, and the like, must be grounded to comply with NEC Article 250.

5-3 On a switchboard with busbars, phase B would have the highest voltage to ground in a:

A) Delta-Wye system

B) High-leg delta-connected system

C) Wye-Wye system

D) Scott connection

Answer: B

See Figure 5-1.

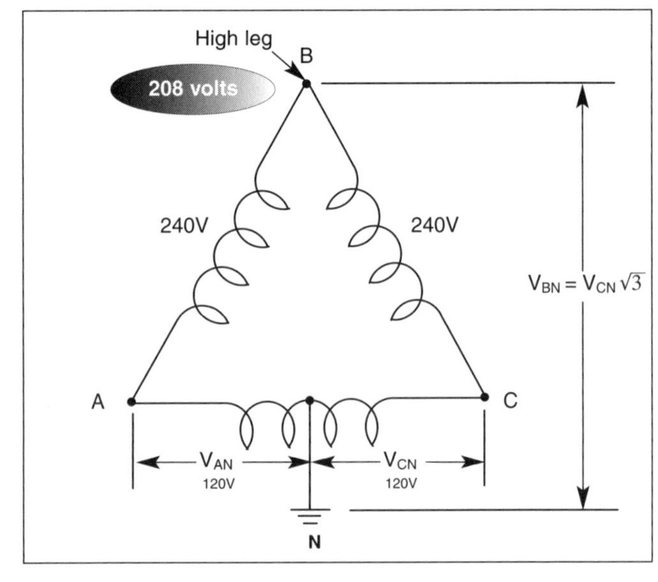

Figure 5-1: In a three-phase, 4-wire, delta-connected transformer with a grounded center tap, the voltage from phase B to neutral is higher than either phase A or phase C to ground. Multiply the voltage of either phase A or C to ground by the square root of 3. Therefore, 120 x 1.73 = 207.6 volts from phase B to ground. This rounds off to 208 volts.

5-4 A 200-ampere panelboard may supply a maximum load of:

A) 100 amperes
B) 180 amperes
C) 160 amperes
D) 200 amperes

Answer: D

A panelboard may supply a load not in excess of its rating. NEC Section 408.30.

5-5 A lighting and appliance branch-circuit panelboard has 10% or more of its overcurrent devices rated:

A) 30 amperes or less
B) 25 amperes or less
C) 20 amperes or less
D) 15 amperes or less

Answer: A

A lighting and appliance branch-circuit panelboard is one having more than 10% of its overcurrent devices rated 30 amperes or less, for which branch-circuit grounded connections are provided. NEC Section 408.34.

5-6 The terminal bar in a panelboard is connected to the neutral bar only when the panelboard is used as:

A) Motor control feeders
B) Low-voltage feeders
C) Service equipment
D) Machinery feeders

Answer: C

The terminal bar used for equipment grounding (bonding) in a panelboard may be connected together only when used as service equipment. When used otherwise, the terminal bar and neutral bar must be isolated from each other. NEC Section 250.24(A).

5-7 In a data-processing room the disconnecting means must be at:

A) Each group of computers
B) The principal exit door
C) Each piece of electronic equipment
D) The main service entrance

Answer: B

The disconnect switches for electrical feeders in a computer or data-processing room must be located at the principal exit door to deenergize all equipment upon exiting the room. NEC Section 645.10 now allows the disconnecting means to be in the form of a pushbutton.

5-8 What is the maximum number of overcurrent devices, excluding the mains, allowed in a lighting and appliance panelboard?

A) 38

B) 40

C) 42

D) 44

Answer: C

The maximum number of overcurrent devices in a three-phase, 4-wire system is 42. NEC Section 408.35.

5-9 Insulated grounded conductors, in sizes larger than No. 6 AWG, may be identified by:

A) A continuous white or natural gray outer finish

B) Three continuous white stripes on other than green insulation

C) Distinctive white markings encircling the conductor at its terminations

D) All of the above

Answer: D

NEC Section 200.6(B) permits grounded conductors to be identified by a white or gray outer finish, or by three continuous white stripes along their entire length on other than green insulation. However, only sizes larger than No. 6 AWG are permitted to be marked as white or gray at points of termination, and such markings must encircle the entire conductor.

5-10 The minimum size copper equipment grounding conductor for a 200-ampere circuit is:

A) No. 3 AWG

B) No. 6 AWG

C) No. 8 AWG

D) No. 10 AWG

Answer: B

NEC Table 250.122 gives No. 6 AWG copper wire as the minimum size for a 200-ampere circuit.

5-11 Auxiliary gutters must not be filled to greater than:

A) 20%

B) 30%

C) 40%

D) 50%

Answer: A

NEC Section 366.22 requires that the sum of the cross-sectional areas of all contained conductors at any cross section of an auxiliary gutter shall not exceed 20% of the interior cross-sectional area of the auxiliary gutter.

5-12 The working clearance between live parts for a system operating at 480 volts with exposed live parts on one side and grounded parts on the other side is:

A) 1 foot

B) 2 feet

C) 3.5 feet

D) 4 feet

Answer: C

NEC Table 110.26(A) lists this situation as condition 2 and specifies a minimum of 3½ feet clearance.

5-13 A 200-ampere lighting and appliance branch-circuit panelboard may be protected by two main breakers with a combined rating of:

A) 150 amperes

B) 175 amperes

C) 200 amperes

D) 225 amperes

Answer: C

The combined overcurrent protection must not exceed the rating of the panelboard. NEC Section 408.36(A).

5-14 Panelboards equipped with 30-ampere or less snap switches must be protected not in excess of:

A) 100 amperes

B) 150 amperes

C) 200 amperes

D) 225 amperes

Answer: C

The maximum rating of panelboards containing snap switches with ratings of 30 amperes or less is 200 amperes according to NEC Section 408.36(C).

5-15 Panelboards installed in health care facilities and serving the same individual patient vicinity must be bonded together with copper conductors not smaller than:

A) No. 12 AWG

B) No. 10 AWG

C) No. 8 AWG

D) No. 6 AWG

Answer: B

The equipment grounding terminal busses of branch-circuit panelboards, serving the same individual patient vicinity, shall be bonded together with an insulated continuous copper conductor not smaller than No. 10 AWG. NEC Section 517.14.

5-16 The panels of switchboards must be made of:

A) Moisture-resistant, noncombustible material

B) Raintight material

C) Impact-resistant material

D) Waterproof material

Answer: A

All panels of switchboards must be moisture resistant and composed of noncombustible material. If the panel is used in other than conventional atmospheres, the housing must be modified to suit the condition. NEC Section 408.50.

5-17 How shall busbars be mounted in a panelboard?

A) Snug, but not tight

B) Loose

C) Rigidly

D) Horizontal to the floor

Answer: C

Insulated or bare busbars shall be rigidly mounted. NEC Section 408.51.

5-18 The maximum rating of overcurrent devices protecting panelboard instrument circuits is:

A) 10 amperes

B) 20 amperes

C) 15 amperes

D) 30 amperes

Answer: C

Instrument circuits such as pilot lights and potential transformers shall be supplied by a circuit that is protected by standard overcurrent devices rated 15 amperes or less. NEC Section 408.52.

5-19 When blades of knife switches are open, they must be:

A) Energized

B) De-energized

C) Warning tags installed

D) Insulated

Answer: B

NEC Section 404.6(C). See Figure 5-2. The NEC has been revised to require bolted pressure contact switches to have barriers to prevent inadvertent contact with the energized blades.

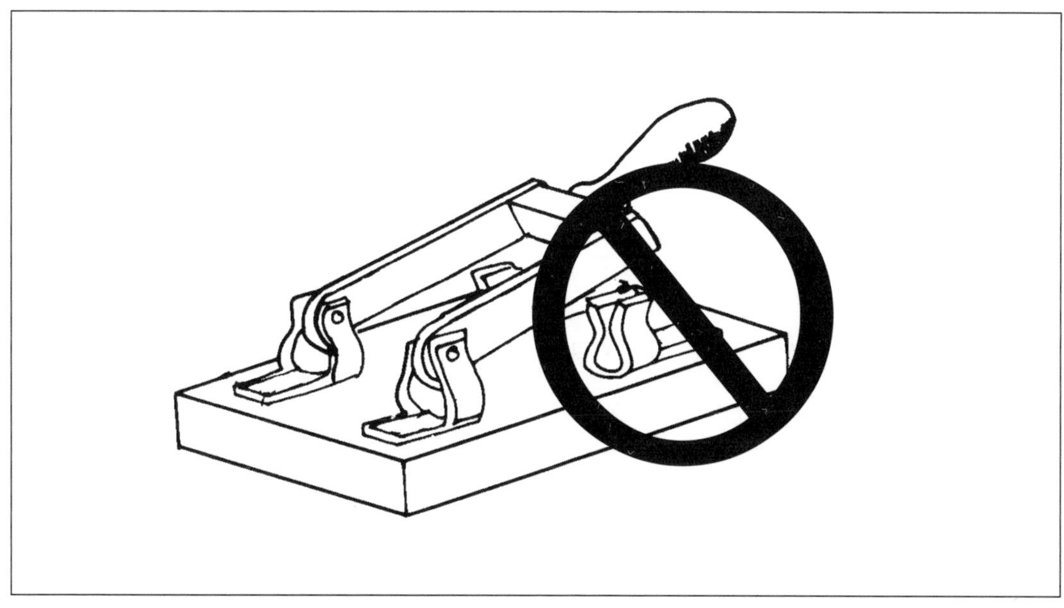

Figure 5-2: A knife switch must be arranged so it will not close by gravity. The blades, when opened, must be "dead" or de-energized. Therefore, the line must be connected at the top of the switch and the load connected at the bottom where the blade hinges.

5-20 What is the minimum wire bending space allowed in a panelboard being fed with paralleled pairs of 2/0 conductors entering the enclosure through the wall opposite the terminals?

A) 6 inches

B) 7 inches

C) 8 inches

D) 9 inches

Answer: A

NEC Table 312.6(B).

5-21 What is the minimum length required of a junction box in a straight pull?

A) Eight times the cross sectional area of the smallest conductor

B) Four times the cross sectional area of the largest conductor

C) Eight times the trade diameter of the largest raceway

D) Six times the trade diameter of the largest raceway

Answer: C

In straight pulls, the length of the junction box must be at least eight times the diameter of the largest raceway. NEC Section 314.28(A)(1).

5-22 Knife switches rated at over 1200 amperes at 250 volts or less, and at over 600 amperes at 251 to 600 volts, shall be opened only under what condition?

A) When switches are fully loaded

B) When switches are isolated

C) In case of fire

D) Under no load

Answer: D

Knife switches as specified in this question must not be opened under load. Therefore, if a knife switch is providing a disconnecting means for a feeder serving, say, six motor controllers, each motor must be shut off before opening the knife switch, or else enough motors must be shut off to obtain conditions below 1200 amperes at 250 volts or below 600 amperes at from 251 to 600 volts. NEC Section 404.13(A).

5-23 Panelboards must be mounted in:

A) Cabinets

B) Enclosures designed for the purpose

C) Cutout boxes

D) Any of these

Answer: D

NEC Section 408.38 allows panelboards to be enclosed in either cabinets, cutout boxes, or enclosures designed for the purpose.

5-24 What type of front must be used for panelboard enclosures?

A) Live front

B) Dead front

C) Gray-colored front

D) Gray-colored front with yellow stripes

Answer: B

NEC Section 408.38 requires all panelboard enclosures to be of the dead front type.

5-25 Metal panelboard cabinets and their related frames must be:

A) Isolated from each other

B) Separately bonded

C) In physical contact with each other

D) None of the above

Answer: C

NEC Section 408.40 requires panelboard cabinets and their related frames, if of metal, to be in physical contact with each other and grounded in accordance with NEC Article 250.

Panel No	Type Cabinet	Panel Mains			Branches					Items Fed
		Amps	Volts	Phase	1P	2P	3P	Prot.	Frame	
"A"	Flush	200 A	120/240V	3∅, 4W Y	—	1	—	20A	70A	Time clock
					—	—	1	20A	70A	A.H.U
					—	1	—	30A	70A	Water htr.
					—	—	1	30A	70A	Cond. unit
					5	—	—	20A	70A	Lights
					10	—	—	20A	70A	Recepts.
					5	—	—	20A	70A	Spares
					12	—	—	—	—	Provisions only

Figure 5-3: Typical panelboard schedule used on an electrical drawing

5-26 The panelboard schedule in Figure 5-3 describes the characteristics of an electric panelboard. To which panel number or letter does the schedule refer?

A) A

B) B

C) C

D) D

Answer: A

In the top left-hand corner of the schedule is a column heading designated "Panel No." and under this is the designation "A."

5-27 Referring again to Figure 5-3, what type of cabinet is used?

A) Flush-mounted

B) PVC plastic

C) Surface-mounted

D) 10 gauge sheet metal

Answer: A

Under the heading "Type Cabinet" in the schedule is the word "Flush."

5-28 How many one-pole circuit breakers are allocated for lighting in the panelboard schedule in Figure 5-3?

A) 12
B) 10
C) 7
D) 5

Answer: D

Look under the major heading "Branches" and then look under the subheading "1P" to see that "5" 20A circuit breakers are used for lights.

5-29 What is the protection rating of the circuit breaker used for the water heater in Figure 5-3?

A) 15 amperes
B) 20 amperes
C) 30 amperes
D) 40 amperes

Answer: C

Look in the right-hand column under "Items Fed" to see that the water heater (htr.) is protected with a 2-pole circuit breaker with "30A" (ampere) protection (Prot.) rating.

5-30 How many "Provisions only" are allocated in the panelboard schedule in Figure 5-3?

A) 10
B) 12
C) 14
D) 16

Answer: B

Look under "Items Fed" in the right-hand column heading, then down to "Provisions only" in this column. Looking to the left, you can see that "12" 1P spaces are allocated.

5-31 The working clearance between live parts for a system operating at 480 volts with exposed live parts on one side and no live or grounded parts on the other side of the working space must be a minimum of:

A) 3 feet
B) 3.5 feet
C) 4 feet
D) 5 feet

Answer: A

NEC Table 110.26(A) lists this situation as Condition 1 and calls for 3 feet working clearance.

5-32 What size and type of continuous conductor is required for a bonding conductor between equipment grounding terminal busses of normal and essential branch-circuit panelboards in health care facilities?

A) No. 10 AWG insulated aluminum

B) No. 12 AWG insulated copper

C) No. 10 AWG insulated copper

D) No. 12 AWG insulated aluminum

Answer: C

Such panelboards serving the same individual patient area must be bonded together with an insulated continuous copper conductor not smaller than No. 10 AWG. NEC Section 517.14.

5-33 What is the allowable ampacity for a 1/0 THHN conductor in a conduit with 5 other current carrying conductors?

A) 170 amps

B) 150 amps

C) 136 amps

D) 200 amps

Answer: C

NEC Table 310.16 lists the ampacity for a 1/0 THHN conductor at 170 amps. However, Article 310.15 requires conductors to be derated when more than 3 current carrying conductors are in the same cable or raceway. A 1/0 THHN conductor at 80% would be rated at 136 amperes.

5-34 What must be provided in a lighting and appliance panelboard to prevent installing more overcurrent devices than the number for which the panelboard is rated?

A) A warning sign

B) A highly visible tag

C) A padlock

D) Physical means

Answer: D

NEC Section 408.35 requires such panelboards to be provided with a physical means to prevent the installation of more overcurrent devices. One method is to provide a certain length of busbar to limit the amount of circuit breakers that can be attached.

5-35 When keeping within the NEC requirements for the number of overcurrent devices allowed in a lighting and appliance panelboard, how many overcurrent devices are counted in one 2-pole circuit breaker?

A) 1

B) 2

C) 3

D) 4

Answer: B

Since a 2-pole circuit breaker takes up two spaces in the panelboard, it is counted as two overcurrent devices. NEC Section 408.35.

5-36 How many overcurrent device spaces are counted when using one 3-pole circuit breaker?

A) 1

B) 2

C) 3

D) 4

Answer: C

Since a 3-pole circuit breaker takes up three spaces in a panelboard, it is counted as three overcurrent devices. NEC Section 408.35.

5-37 A 40-circuit, single-phase, 3-wire lighting and appliance panelboard contains a 200-ampere 2-pole main circuit breaker, and an additional 18 1-pole circuit breakers. What is the maximum number of 2-pole circuit breakers that may be installed in the panelboard?

A) 10

B) 11

C) 12

D) 13

Answer: B

The 200-ampere 2-pole main circuit breaker doesn't count toward the maximum allowable number (NEC Section 408.35). Forty 1-pole spaces are available in this panelboard. If 18 of these spaces are taken up with the 18 1-pole breakers, this leaves (40 − 18 =) 22 spaces available. Since a 2-pole breaker takes up two spaces, the panelboard will accommodate 11 additional 2-pole circuit breakers (22 divided by 2 equals 11 circuit breakers).

5-38 What is the maximum number of main circuit breakers that may be used to protect a lighting and appliance panelboard on the supply side?

A) None

B) 1

C) 2

D) 3

Answer: C

NEC Section 408.36(A) limits the number to two main circuit breakers or two sets of fuses having a combined rating not greater than that of the panelboard.

5-39 When is it permissible to install two wires under the same set-screw in a neutral?

A) Only on branch circuits

B) Only on relays mounted on switchboards

C) When the terminal is identified as allowing more than one conductor

D) It is never permitted

Answer: C

NEC Section 110.14(A) states that terminals for more than one conductor shall be so identified.

5-40 A metal pole supporting a light fixture shall have an accessible handhole not less than ______ inches.

A) 2 × 6 inches

B) 3 × 4 inches

C) 2 × 4 inches

D) 4 × 8 inches

Answer: C

NEC Section 410.15(B) requires handholes to be 2 × 4 inch minimum. In addition, the cover must be raintight.

5-41 What is the maximum overcurrent device rating allowed for the protection of panelboard instrument circuits?

A) 30 amperes

B) 20 amperes

C) 15 amperes

D) 10 amperes

Answer: C

NEC Section 408.52 limits the overcurrent device to 15 amperes or less.

5-42 What is the minimum wire bending space required at the top and bottom of a panelboard that has one 3/0 conductor connected to each busbar in the panelboard?

A) $6\frac{1}{2}$ inches

B) 7 inches

C) $7\frac{1}{2}$ inches

D) 8 inches

Answer: A

NEC Table 312.6(B) requires the minimum space to be $6\frac{1}{2}$ inches.

5-43 What is the minimum wire bending space required at the top and bottom of a panelboard that has two 3/0 conductors connected to each busbar or terminal in the panelboard?

A) $6\frac{1}{2}$ inches

B) 7 inches

C) $7\frac{1}{2}$ inches

D) 8 inches

Answer: A

NEC Table 312.6(B) requires the minimum space to be $6\frac{1}{2}$ inches.

5-44 What is the minimum wire bending space required at the top and bottom of a panelboard that has three 3/0 conductors connected to each busbar or terminal in the panelboard?

A) $6\frac{1}{2}$ inches

B) 7 inches

C) $7\frac{1}{2}$ inches

D) 8 inches

Answer: D

NEC Table 312.6(B) requires the minimum space to be 8 inches when three 3/0 conductors are connected to one terminal.

5-45 All switches and circuit breakers, used as switches under normal operation, must be so located that they may be operated from:

A) A cat walk

B) A power pole

C) An inaccessible location

D) A readily accessible place

Answer: D

NEC Section 404.8(A) requires all switches and circuit breakers used as switches to be installed so they may be operated (turned on and off) from a readily accessible place.

5-46 What is the highest position the operating handle of a safety switch or circuit breaker may be located above the floor or working platform under normal working conditions?

A) 4 1/2 feet

B) 5 feet

C) 6 feet

D) 6 feet 7 inches

Answer: D

For normal safety switch and circuit breaker operation, NEC Section 404.8(A) requires the maximum height to be 6 feet 7 inches.

5-47 On busway installations, fused switches and circuit breakers may be located at the same level as the busway provided:

A) A suitable means is provided to operate the device from a step ladder

B) A suitable means is provided to operate the device from floor level

C) A suitable means is provided to operate the device from an extension ladder

D) Another disconnect switch is provided at the working platform

Answer: B

NEC Section 404.8(A), Exception 1 requires that a suitable means must be provided to operate the fused switch or circuit breaker from floor level.

5-48 When installing a grounded conductor in a panelboard each grounded conductor must be terminated in an individual terminal that is not also used for another conductor unless:

A) Terminal is identified for use with more than one conductor

B) Conductors are under No. 2 AWG

C) Conductors are under No. 4 AWG

D) Not permitted under any situation

Answer: A

NEC Section 408.41 explains the requirements for terminating the grounded conductor in a panel. It also permits more than one conductor to be placed in the same terminal if the terminal is identified for such use.

5-49 What is the maximum voltage allowed between adjacent snap switches, receptacles, and similar devices in an enclosure?

A) 120 volts
B) 240 volts
C) 300 volts
D) 480 volts

Answer: C

NEC Section 404.8(B) limits the voltage to 300 unless they are installed in enclosures equipped with permanently installed barriers between adjacent switches. The code has been revised to include receptacles and similar devices.

5-50 If safety switches are horsepower rated, the rating must be marked on the switch as well as:

A) Current and voltage rating
B) Voltage and wattage rating
C) Current and resistance rating
D) Voltage and resistance rating

Answer: A

NEC Section 404.15(A) requires switches to be marked with their current and voltage rating.

5-51 A fused safety switch must not have the fuses connected in:

A) Series
B) Parallel
C) Line with the load
D) None of the above

Answer: B

NEC Section 404.17 requires that fused switches have their fuses connected in series with the load.

5-52 A cutout box installed in a wet location shall be:

A) Raintight
B) Waterproof
C) Weatherproof
D) Rainproof

Answer: C

NEC Section 312.2(A) requires the enclosure to be weatherproof. In addition, a 1/4-inch airspace between the box and the wall is required.

5-53 The minimum spacing allowed between bare metal surface-mounted, current-carrying parts of opposite polarity in a panelboard with voltage not exceeding 125 volts is:

A) $^{1}/_{4}$ inch

B) $^{1}/_{2}$ inch

C) $^{3}/_{4}$ inch

D) 1 inch

Answer: C

NEC Table 408.56 specifies $^{3}/_{4}$ inch for this condition.

5-54 If the bare metal parts specified in Question 5-53 are held free in air, what is the minimum distance?

A) $^{1}/_{4}$ inch

B) $^{1}/_{2}$ inch

C) $^{3}/_{4}$ inch

D) 1 inch

Answer: B

NEC Table 408.56 specifies $^{1}/_{2}$ inch for this condition.

5-55 The minimum spacing allowed between bare metal surface-mounted, current-carrying parts to ground in a panelboard with voltage not exceeding 125 volts is:

A) $^{1}/_{4}$ inch

B) $^{1}/_{2}$ inch

C) $^{3}/_{4}$ inch

D) 1 inch

Answer: B

NEC Table 408.56 specifies $^{1}/_{2}$ inch for this condition.

5-56 The minimum spacing allowed between bare metal surface-mounted, current-carrying parts of opposite polarity in a panelboard with voltage not exceeding 250 volts is:

A) $^{1}/_{2}$ inch

B) $^{3}/_{4}$ inch

C) 1 inch

D) $1^{1}/_{4}$ inches

Answer: D

NEC Table 408.56 specifies $1^{1}/_{4}$ inches for this condition.

5-57 The minimum spacing allowed between bare metal current-carrying parts (not exceeding 250 volts) of opposite polarity where held in free air is:

A) 1/2 inch

B) 3/4 inch

C) 1 inch

D) 1 1/4 inches

Answer: B

NEC Table 408.56 requires 3/4 inch minimum for this condition.

5-58 The minimum spacing allowed between bare metal current-carrying parts to ground in a panelboard with voltage not exceeding 250 volts is:

A) 1/2 inch

B) 3/4 inch

C) 1 inch

D) 1 1/4 inches

Answer: A

NEC Table 408.56 requires 1/2 inch minimum spacing for this condition.

5-59 A lighting and appliance panelboard has what percentage of its circuits supplying lighting and appliance loads rated 30 amps or less with neutral connected loads?

A) 20%

B) 30%

C) 5%

D) 10%

Answer: D

NEC Section 408.34(A) specifies 10%. This article has been rewritten to clarify the definition of a lighting and appliance panelboard.

5-60 The minimum spacing allowed between current-carrying conductors in free air when the voltage does not exceed 600 volts is:

A) 1 inch

B) 2 inches

C) 3 inches

D) 4 inches

Answer: A

NEC Table 408.56 specifies a minimum 1-inch space under this condition.

5-61 The minimum spacing between bare metal 600-volt current-carrying parts and ground is:

A) 1 inch
B) 2 inches
C) 3 inches
D) 4 inches

Answer: A

NEC Table 408.56 specifies 1 inch for this condition.

5-62 May enclosures for switches or overcurrent devices be used as junction boxes?

A) Yes, in all conditions
B) No, in all conditions
C) Under certain conditions
D) None of the above

Answer: C

Where adequate space is provided so that the conductors do not fill the wiring space or any cross section to more than 40% of the cross-sectional area of the space, and the conductors, splices, and taps do not fill the wiring space at any cross section to more than 75% of the cross-sectional area of the space, the enclosure may be used for a junction box. NEC Section 312.8.

5-63 Which of the following must be provided when a No. 4 AWG or larger conductor enters a panelboard?

A) A bonding jumper
B) A ground clip
C) An insulated bushing
D) An identification tag

Answer: C

NEC Section 300.4(F) requires an insulating fitting such as an insulated bushing.

5-64 In Question 5-63 above, what condition would not require an insulating bushing?

A) When the conductors are secured with tie-straps
B) When the conductors are correctly tagged
C) When No. 6 AWG or larger wire is used
D) When the conductors are separated from the fitting by substantial insulating material

Answer: D

NEC Section 300.4(F) does not require an insulating bushing if the conductors are separated from the raceway fitting by substantial insulating material securely fastened in place.

5-65 May conduit bushing constructed wholly of insulating material be used to secure a raceway to a panelboard?

A) Yes

B) No

C) Under certain conditions

D) None of the above

Answer: B

A bushing constructed wholly of insulating material will not adequately secure a raceway system. The raceway must have metal fittings for support and an insulating bushing for conductor protection. NEC Section 300.4(F).

5-66 Where must metal panelboards be protected against corrosion?

A) Outside only

B) Inside only

C) Both inside and outside

D) None of the above

Answer: C

NEC Section 312.10(A) requires that metal cabinets and cutout boxes be protected both inside and outside against corrosion.

Chapter 6

Overcurrent Protection

Reliable overcurrent protective devices, such as fuses and circuit breakers, prevent or minimize costly damage to transformers, conductors, motors, equipment, and the other many components and loads that make up a complete electrical system.

A fuse is the simplest device for opening an electric circuit when excessive current flows because of an overload or such fault conditions as grounds and short circuits. A "fusible" link or links encapsulated in a tube and connected to contact terminals comprise the fundamental elements of the basic fuse. Electrical resistance of the link is so low that it simply acts as a conductor and every fuse is intended to be connected in series with each phase conductor so that current flowing through the conductor to any load must also pass through the fuse. The continuous current rating of a fuse in amperes establishes the maximum amount of current the fuse will carry without opening. When circuit current flow exceeds this value, an internal element (link) in the fuse melts due to the heat of the current flow and opens the circuit. Fuses are manufactured in a wide variety of types and sizes with different current ratings, different abilities to interrupt fault currents, various speeds of operation (either quick-opening or time-delay opening), different internal and external constructions, and voltage ratings for both low-voltage (600 volts and below) and medium-voltage (over 600 volts) circuits.

A circuit breaker resembles an ordinary toggle switch, and it is probably the most widely used means of overcurrent protection today. On an overload, the circuit breaker opens itself or *trips*. In a tripped position, the handle jumps to the middle position as shown in Figure 1-7 in Chapter 1. To reset, turn the handle to the OFF position and then turn it as far as it will go beyond this position; finally, turn it to the ON position.

On a conventional 120/240-volt, single-phase electric service, one single-pole circuit breaker protects one 120-volt circuit, and one double-pole circuit breaker protects one 240-volt circuit. Three-phase electric services will require 3-pole circuit breakers for three-phase, 3- or 4-wire circuits.

Circuit breakers are rated in amperes, just like fuses, although the particular ratings are not exactly the same as those for fuses.

6-1 The highest current rating of an Edison-base plug fuse is:

A) 30 amperes
B) 60 amperes
C) 100 amperes
D) 200 amperes

Answer: A

Plug fuses of the Edison-base type shall be classified at not over 125 volts and 30 amperes. NEC Section 240.51(A).

6-2 Edison-base plug fuses may be used only in:

A) New work
B) Existing installations under specified conditions
C) Motor control panels
D) Motor circuits

Answer: B

Plug fuses of the Edison-base type shall be used only for replacements in existing installations where there is no evidence of overfusing or tampering. NEC Section 240.51(B).

6-3 Which of the following is a nonstandard ampere rating for a plug fuse?

A) 15 amperes
B) 20 amperes
C) 30 amperes
D) 35 amperes

Answer: D

The maximum standard rating for plug fuses is 30 amperes. NEC Section 240.51(A).

6-4 The maximum size fuse allowed to protect a No. 14 AWG THW conductor is:

A) 10 amperes
B) 15 amperes
C) 20 amperes
D) 25 amperes

Answer: B

The maximum current-carrying capacity of No. 14 AWG THW conductors is 15 amperes. Therefore, the fuse size must not exceed that rating. NEC Table 310.16 and Section 240.4(D).

6-5 When a circuit breaker trips, its operating handle will be in which of the following positions?

A) Closed (energized) position

B) Open position

C) Half-way between the ON and OFF positions

D) No change

Answer: C

The conventional circuit-breaker handle has four possible positions: on, off, trip, and reset. When tripped, the handle jumps to the middle position.

6-6 Plug fuses of 15 ampere and lower rating must be identified by what type of window configuration?

A) Round

B) Hexagonal

C) Square

D) Octagonal

Answer: B

Plug fuses of 15 ampere and lower rating shall be identified by a hexagonal configuration of the window, cap, or other prominent part to distinguish them from fuses of higher ampere ratings. NEC Section 240.50(C).

6-7 Which of the following conductors need overcurrent protection on a residential electric service?

A) Grounded conductor

B) Bonding conductor

C) Ungrounded conductors

D) Equipment grounding conductor

Answer: C

A fuse or an overcurrent trip unit of a circuit breaker shall be connected in series with each ungrounded conductor. NEC Section 240.20(A).

6-8 A circuit feeding three single-phase motors with full-load current ratings of 40, 20 and 20 amperes, respectively, requires the following dual-element time-delay fuse size:

A) 110 amperes

B) 150 amperes

C) 200 amperes

D) 250 amperes

Answer: A

Size the fuse at 175% of the full-load current of the largest motor plus the full-load current of all other motors. NEC Section 430.53(C)(4).

6-9 A nontime-delay fuse protects a circuit with all motor loads with the largest motor drawing a full-load current of 40 amperes, and two other motors drawing a full-load current of 15 amperes each. What fuse size should be used?

A) 100 amperes

B) 150 amperes

C) 175 amperes

D) 250 amperes

Answer: B

Size the fuse at 300% of the full-load current of the largest motor plus the full-load current of all other motors. NEC Section 430.53(C)(4).

6-10 When using a time-delay fuse to protect a circuit with no motor load, the fuse size must be at least:

A) 80% of the continuous load

B) 100% of the continuous load

C) 125% of the continuous load plus 100% of the noncontinuous load

D) 150% of the continuous load plus 125% of the noncontinuous load

Answer: C

The fuse and branch-circuit sizes must be at least 125% of the continuous load plus 100% of the noncontinuous load. NEC Section 210.20(A).

6-11 Conductors may be tapped, without overcurrent protection at the tap, to a feeder or transformer secondary if certain conditions are met. If the ampacity of the tap conductor is not less than the combined computed loads on the circuits supplied by the tap conductor, the length of the tap conductor cannot exceed:

A) 5 feet

B) 10 feet

C) 15 feet

D) 25 feet

Answer: B

The length of the tap conductor must not exceed 10 feet. NEC Section 240.21(B)(1).

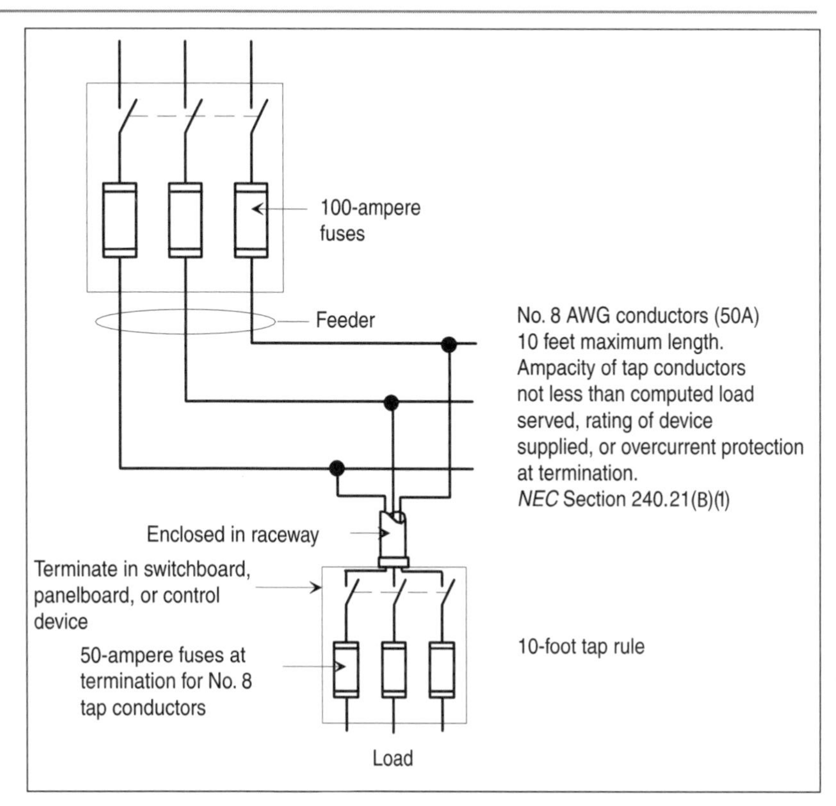

Figure 6-1: Summary of the *NEC* "10-foot" tap rule

6-12 Where must an overcurrent device be located in a circuit?

A) At the point where the conductor receives its supply

B) At the last outlet on the circuit

C) On the line side of the electric meter

D) At the first fitting or connector in the circuit

Answer: A

An overcurrent device shall be connected at the point where the conductor to be protected receives its supply. NEC Section 240.21.

6-13 The fuse in Figure 6-2 is known as:

A) Edison-base plug fuse

B) Type S plug fuse

C) Edison-base cartridge fuse

D) Nonrenewable cartridge fuse

Figure 6-2: One type of plug fuse

Answer: A

The Edison-base plug fuse is the standard screw-in base for fuse holders and is larger in diameter than Type S fuses. Therefore, Type S fuses require an adapter when they are used.

6-14 The fuse in Figure 6-3 is known as:

A) Despard plug fuse

B) Type S plug fuse

C) Dual-element time-delay plug fuse

D) Renewable cartridge fuse

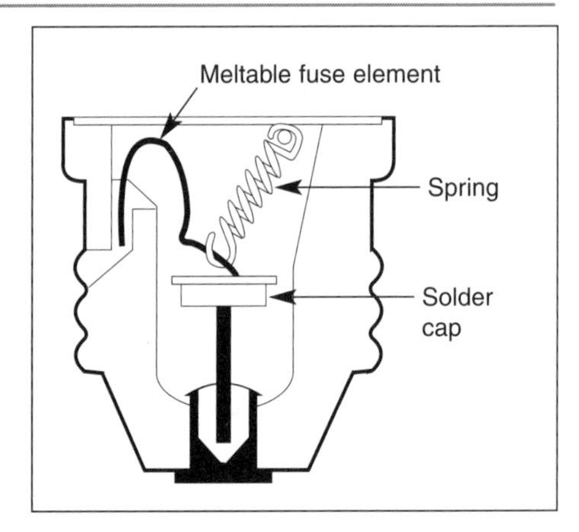

Figure 6-3: One type of plug fuse

Answer: C

The dual-element time-delay plug fuse is recommended for small household motor-operated appliances such as window air conditioners, sump pumps, etc.

6-15 The fuse in Figure 6-4 is known as:

A) Dual-element time-delay cartridge fuse

B) Type S plug fuse

C) Nontime-delay cartridge fuse

D) Nonrenewable cartridge fuse

Answer: C

The nontime-delay cartridge fuse has only one element.

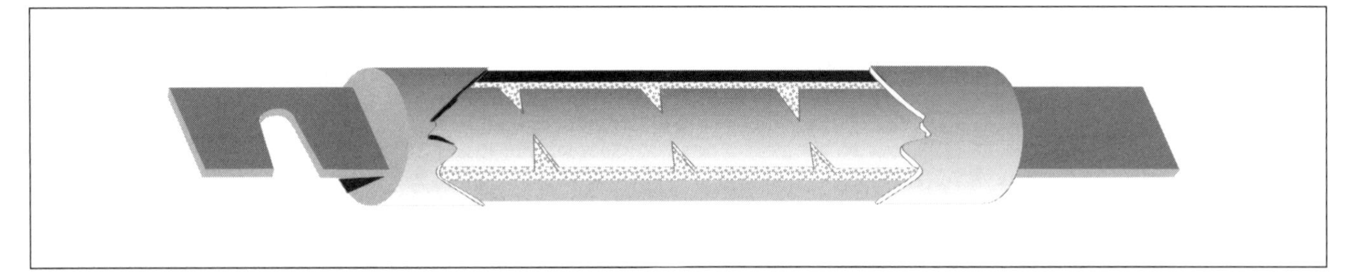

Figure 6-4: One type of cartridge fuse

6-16 The fuse in Figure 6-5 is known as:

A) Dual-element time-delay cartridge fuse

B) Type S plug fuse

C) Type S fuse adapter

D) Nonrenewable cartridge fuse

Answer: A

This type of cartridge fuse has distinct and separate overload and short-circuit elements.

6-17 The fuse clips in Figure 6-6 are known as:

A) Class A fuse rejection clips

B) Type D fuse rejection clips

C) Class R fuse rejection clips

D) Nonrenewable cartridge-fuse clips

Answer: C

Class R fuse clips are designed to prevent older type Class H fuses from being installed.

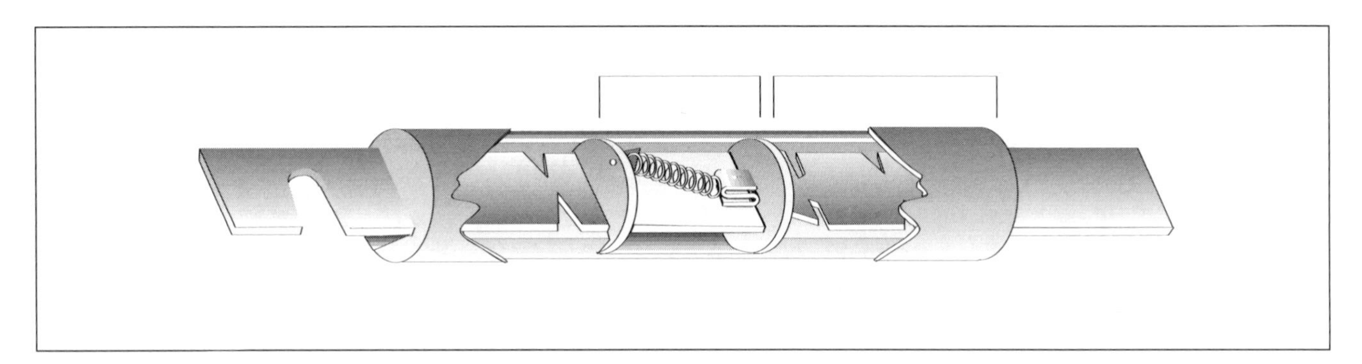

Figure 6-5: One type of cartridge fuse

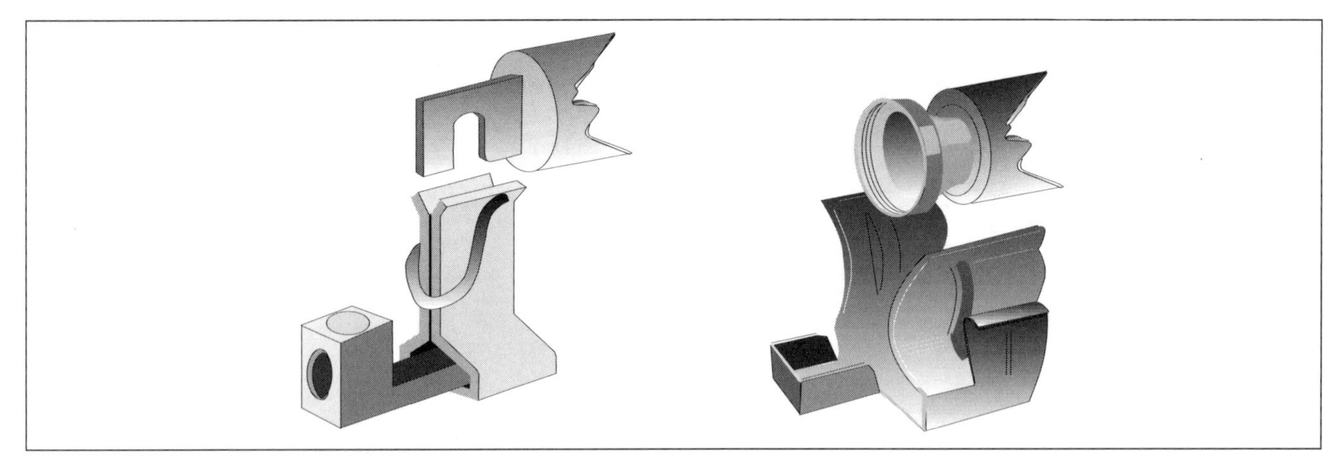

Figure 6-6: Class R fuse rejection clips that accept only Class R fuses

6-18 If a multiwire branch circuit supplies line-to-line loads, the branch-circuit protective device must open each:

A) Ungrounded conductor simultaneously

B) Grounded conductor

C) Ungrounded conductor individually

D) Bonding conductor

Answer: A

For example, if a 3-wire circuit is connected to a 3-pole circuit breaker, and one of the three wires should develop a fault, the circuit breaker must open all three wires simultaneously and not just the one that developed the fault. NEC Section 240.20(B).

6-19 The largest standard rating for fuses and inverse time circuit breakers is:

A) 1000 amperes

B) 1600 amperes

C) 1200 amperes

D) 6000 amperes

Answer: D

The standard ampere ratings for fuses and inverse circuit breakers are 15, 20, 25, 30, 35, 40, 45, 50, 60, 70, 80, 90, 100, 110, 125, 150, 175, 200, 225, 250, 300, 350, 400, 450, 500, 600, 700, 800, 1000, 1200, 1600, 2000, 2500, 3000, 4000, 5000, and 6000 amperes. NEC Section 240-6.

6-20 To meet *NEC* regulations, fuses or circuit breakers connected in parallel must be:

A) Assembled by the electrician or technician on the job

B) Encased in a PVC housing

C) Factory assembled in parallel

D) Rated over 200 amperes

Answer: C

Fuses, circuit breakers, or combinations thereof shall not be connected in parallel unless they are factory assembled. NEC Section 240.8.

6-21 Supplementary overcurrent devices shall:

A) Be required to be readily accessible and within sight of the equipment

B) Not be required to be readily accessible

C) Not be allowed under any conditions

D) Be connected only in parallel or series-parallel configurations

NEC CHANGE

Answer: B

Supplementary overcurrent protection shall not be used as a substitute for required branch-circuit overcurrent devices. Consequently, they need not be readily accessible. NEC Section 240.10.

6-22 One combination that is considered by the *NEC* to be equivalent to an overcurrent trip unit is:

A) Pole-mounted transformer and overcurrent relay

B) Submersible transformer and overcurrent relay

C) Autotransformer and overcurrent relay

D) Current transformer and overcurrent relay

Answer: D

NEC Section 240.20(A) states that a current transformer and overcurrent relay shall be considered equivalent to an overcurrent trip unit.

6-23 When a fuse or circuit breaker is used for circuit protection, the fuse or overcurrent trip unit shall be connected:

A) In series

B) In series and parallel

C) In parallel

D) In tandem

Answer: A

Fuses or overcurrent trip units (circuit breakers) must be connected in series with ungrounded conductors so they will open or trip when an overload or ground fault occurs. NEC Section 240.20(A). The grounding conductor is not provided with overcurrent protection.

6-24 At what point must a branch-circuit overcurrent protection device be placed in an electrical system?

A) At the equipment ground location

B) Where the conductor being protected receives its supply

C) At a tap on the conductor being protected

D) At the outlet where power is consumed

Answer: B

For example, branch circuits must be protected at the panelboard where they receive their supply. NEC Section 240.21.

6-25 The rating of the overcurrent protective device for a circuit supplying a hermetic motor-compressor must be at least what percent of the nameplate rated-load current?

A) 50%

B) 75%

C) 115%

D) 150%

Answer: C

The ampere rating of the overcurrent protective device must be at least 115 percent of the nameplate rated-load current of the compressor or branch-circuit selection current, whichever is greater. NEC Section 440.12(A)(1).

6-26 A multiwire branch circuit protected by fuses may supply only:

A) Half line to line loads and half line to neutral loads

B) Half line to neutral loads to full or trip line to other loads

C) Line to line loads

D) Line to neutral loads

Answer: D

NEC Section 210.4(C) requires that multiwire branch circuits supply only line to neutral loads.

6-27 What is the minimum interrupting rating of an unmarked branch-circuit fuse?

A) 5,000 amperes

B) 10,000 amperes

C) 15,000 amperes

D) 20,000 amperes

Answer: B

NEMA and UL require fuses to be plainly marked with their characteristics including the interrupting rating where it is other than 10,000 amperes. NEC Section 240.60(C).

6-28 A feeder tap less than 25 feet long does not require overcurrent protection at the tap if the ampacity of the tap conductor is at least:

A) 50% of the feeder conductor

B) 40% of the feeder conductor

C) $33^1/_3$% of the feeder conductor

D) 20% of the feeder conductor

Answer: C

NEC Section 240.21(B)(2) requires that feeder tap conductors must not be over 25 feet in length and the ampacity of the tap conductor must be not less than one-third that of the feeder conductor. Therefore, answer C ($33^1/_3$%) is the correct answer. See Figure 6-7 on the next page.

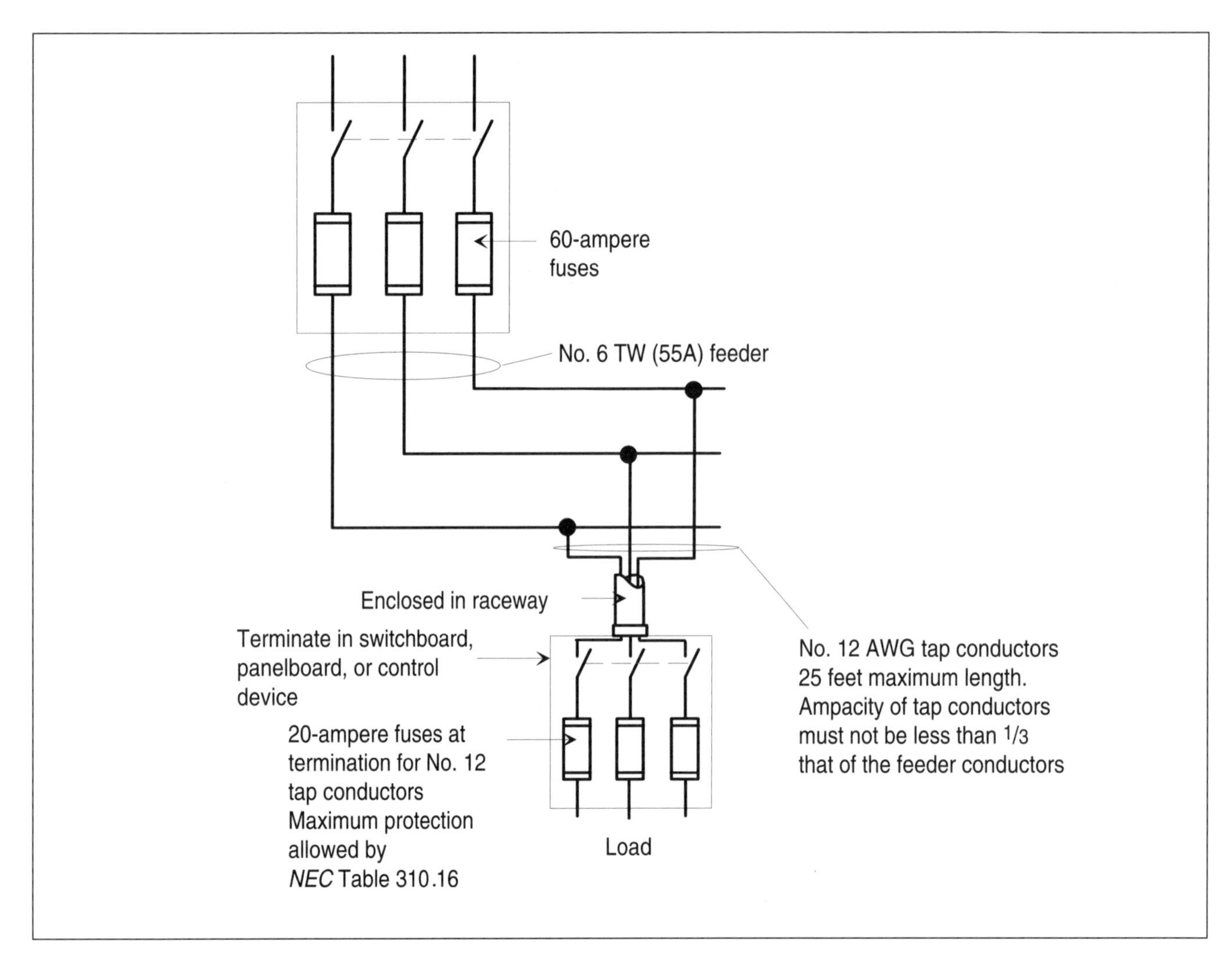

Figure 6-7: Summary of *NEC* 25-foot tap rule

6-29 When designing or installing circuits for diagnostic equipment in health care facilities, what is the minimum percentage allowed for the momentary ampacity of the equipment when sizing branch-circuit conductors and the overcurrent protection?

A) 25%

B) 50%

C) 75%

D) 100%

Answer: B

NEC Section 517.73(A)(1) requires that the ampacity of supply branch-circuit conductors and the current rating of overcurrent protective devices must not be less than 50 percent of the momentary rating or 100 percent of the long-time rating, whichever is greater.

6-30 The rated ampacity of conductors and overcurrent devices of a feeder for two or more branch circuits supplying X-ray equipment for non-medical or non-dental use must not be less than 100% of the momentary demand rating of the two largest pieces of X-ray equipment plus what percent of the momentary ratings of other X-ray apparatus?

A) 5%

B) 10%

C) 15%

D) 20%

Answer: D

NEC Section 660.6(B) specifies 20%. The minimum conductor size for branch and feeder circuits is also governed by voltage regulation requirements. For a specific installation, the manufacturer usually specifies: minimum distribution transformer and conductor sizes, rating of disconnect means, and overcurrent protection.

6-31 What is the maximum overcurrent device rating for each resistance welder connected to an electrical system?

A) Not over 100% of the rated primary current of the welder

B) Not over 200% of the rated primary current of the welder

C) Not over 300% of the rated primary current of the welder

D) Not over 400% of the rated primary current of the welder

Answer: C

NEC Section 630.32(A) specifies a maximum of 300%.

6-32 What is the maximum overcurrent device rating that may be used for each AC transformer or DC rectifier arc welder?

A) Not over 100% of the rated primary current of the welder

B) Not over 200% of the rated primary current of the welder

C) Not over 300% of the rated primary current of the welder

D) Not over 400% of the rated primary current of the welder

Answer: B

NEC Section 630.12(A) restricts the percentage to 200% for this application.

6-33 Each autotransformer 600 volts, nominal, or less shall be protected by an individual overcurrent device. How must this device be connected in relation to the autotransformer?

A) In parallel with each ungrounded input conductor

B) In parallel with each grounded conductor

C) In series with each ungrounded input conductor

D) In series with each grounded conductor

Answer: C

NEC Section 450.4(A) requires overcurrent protection be installed in series with each ungrounded input conductor.

6-34 Such overcurrent devices as described in Question 6-33 must have a rating not exceeding a certain percentage of the rated full-load input current of the autotransformer. What is this percentage?

A) 110%

B) 125%

C) 150%

D) 200%

Answer: B

NEC Section 450.4 limits the overcurrent device rating to 125% of the rated full-load input current of the autotransformer.

6-35 If an autotransformer has a full-load input current of 40 amperes, what is the maximum overcurrent protection device rating allowed?

A) 45 amperes

B) 50 amperes

C) 60 amperes

D) 70 amperes

Answer: B

40 amperes × 1.25 = 50 amperes. NEC Section 450.4.

6-36 In an autotransformer, overcurrent devices shall *not* be installed:

A) In parallel with the output conductors

B) In parallel with the shunt winding

C) In series with the output conductors

D) In series with the shunt winding

Answer: D

NEC Section 450.4(A) requires an overcurrent device to be installed in series with each ungrounded input conductor. However, an overcurrent device shall not be installed in series with the shunt winding.

6-37 A grounding autotransformer is used to create a three-phase, 4-wire distribution system from a three-phase, 3-wire ungrounded system. At what percent must the overcurrent device be set to trip in case of an overload?

A) 100% of the transformer's continuous current per phase or neutral rating

B) 110% of the transformer's continuous current per phase or neutral rating

C) 125% of the transformer's continuous current per phase or neutral rating

D) 150% of the transformer's continuous current per phase or neutral rating

Answer: C

NEC Section 450.5(A)(2) requires the overcurrent device to open or trip if the load on the autotransformer reaches or exceeds 125% of its continuous current per phase or neutral rating. Delayed tripping for temporary overcurrents sensed at the autotransformer overcurrent device is permitted for the purpose of allowing proper operation of branch or feeder protective devices on a 4-wire system.

6-38 What must the continuous neutral current rating be for an autotransformer used to create a three-phase, 4-wire distribution system from a three-phase, 3-wire ungrounded system?

A) It must be sufficient to handle the minimum neutral unbalanced load

B) It must be sufficient to handle the maximum possible neutral unbalanced load

C) It must be sufficient to handle 75% of the neutral unbalanced load

D) It must be sufficient to handle 50% of the unbalanced load

Answer: B

NEC Section 450.5(A)(4) requires the autotransformer to have a continuous neutral current rating sufficient to handle the maximum possible neutral unbalanced load current of the 4-wire system.

6-39 What is the maximum overcurrent rating that may be used for dimmer controls for stage lighting in theaters?

A) 100% of the dimmer rating

B) 110% of the dimmer rating

C) 125% of the dimmer rating

D) 150% of the dimmer rating

Answer: C

NEC Section 520.25(A) limits the overcurrent protection to 125% or less of the dimmer rating.

6-40 If a dimmer control is rated at 2000 watts on a 120-volt single-phase circuit, what is the maximum overcurrent protection allowed for this dimmer?

A) 15 amperes

B) 20 amperes

C) 30 amperes

D) 40 amperes

Answer: B

Using Ohm's law, 2000/120 = 16.666; 16.666 amperes × 1.25 = 20.8 amperes. Therefore, a 20-ampere fuse or circuit breaker is the maximum allowed to comply with NEC Section 520.25(A). A fuse rated for exactly 20.8 amperes is not standard.

6-41 What is the maximum overcurrent protection allowed on circuits feeding stage equipment such as footlights, border lights, etc?

A) 15 amperes

B) 20 amperes

C) 30 amperes

D) 40 amperes

Answer: B

NEC Section 520.41(A) limits the branch circuits feeding such equipment to a maximum load of 20 amperes. Therefore, the overcurrent device (fuse or circuit breaker) must not exceed 20 amperes for circuits feeding stage equipment.

6-42 What is the minimum branch circuit and maximum overcurrent protection rating for a circuit supplying a household electric range rated at 8¾ kW or more?

A) 20 amperes

B) 30 amperes

C) 40 amperes

D) 50 amperes

Answer: C

NEC Section 210.19(A)(3) states that the minimum branch-circuit rating for this load should be 40 amperes. A 40-ampere circuit breaker or cartridge fuse is the maximum allowable rating for a 40-ampere circuit.

6-43 What is the minimum number of circuits allowed to feed a household-type appliance with surface heating elements having a maximum demand of more than 60 amperes?

A) 1

B) 2

C) 3

D) 4

Answer: B

NEC Section 422.11(B) requires that the power supply for such appliances be subdivided into two or more circuits.

6-44 What is the maximum overcurrent protection allowed for each of the circuits described in Question 6-43?

A) 30 amperes

B) 40 amperes

C) 50 amperes

D) 60 amperes

Answer: C

NEC Section 422.11(B) limits the rating to 50 amperes for each subdivided circuit.

6-45 What is the maximum rating of overcurrent devices used to protect circuits feeding infrared lamp heating appliances in commercial and industrial applications?

A) 30 amperes

B) 40 amperes

C) 50 amperes

D) 60 amperes

Answer: C

NEC Section 422.11(C) limits the current rating to 50 amperes.

6-46 How must direct-current conductors used in electroplating systems be protected?

A) Fuses or circuit breakers

B) A current-sensing device that operates a disconnecting means

C) Other approved means by the authority having jurisdiction

D) Any of the above

Answer: D

NEC Section 669.9 recognizes answers A, B, or C as suitable means of protecting dc conductors used in electroplating systems.

6-47 Where the allowable current rating of a busway does not correspond to a standard rating of overcurrent device, what rating is permitted?

A) The next lower rating

B) No overcurrent protection is required

C) The next higher rating

D) None of the above

Answer: C

NEC Section 240.4(B) permits the next higher rating of overcurrent device to be used when the busway is not a standard rating and it is operating at 800 amps or less.

6-48 If the allowable current rating of a busway is 46 amperes, what is the maximum allowable rating of the overcurrent device?

A) 40 amperes

B) 50 amperes

C) 60 amperes

D) 70 amperes

Answer: B

A rating of 46 amperes is not a standard overcurrent device rating. However, 50 amperes is a standard rating and may be used in accordance with NEC Section 240.4(B).

6-49 Omission of overcurrent protection is permitted at points where busways are reduced in size, provided that the smaller busway length does not exceed what distance?

A) 10 feet

B) 20 feet

C) 30 feet

D) 50 feet

Answer: D

NEC Section 368.17(B), Exception, limits the length of the smaller busway to 50 feet. This exception applies only to busways in industrial locations.

6-50 In order to comply with *NEC* regulations what must the current rating of the smaller busway be in Question 6-49?

A) One-third the rating or setting of the overcurrent device next back on the line

B) Half the rating of the overcurrent device next back on the line

C) Three-fourths the rating of the overcurrent device next back on the line

D) Two-thirds the rating of the overcurrent device next back on the line

Answer: A

NEC Section 368.17(B), Exception, requires the smaller busway to have a rating of not less than one-third the rating of the overcurrent device next back on the line. Furthermore, the smaller busway must not be in contact with any combustible material. This applies only to busways in industrial locations.

6-51 What type of overcurrent device is required on a busway system where a branch circuit or subfeeder is tapped from the main busway?

A) Internally operable circuit breaker

B) Internally operable fusible switch

C) Externally operable circuit breaker or externally operable fusible switch

D) None is required

Answer: C

Either an externally operable circuit breaker or fusible switch is permitted in NEC Section 368.17(C).

6-52 When the overcurrent devices in Question 6-51 are mounted out of reach and contain disconnecting means, what provisions must be provided for operating the disconnecting means from the floor?

A) Ropes

B) Chains

C) Sticks

D) Any of the above

Answer: D

NEC Section 368.17(C) allows any of these to serve as a suitable means to reach the disconnecting handle from the floor.

6-53 When the allowable ampacity of a cablebus system does not correspond to a standard rating of overcurrent protection devices, what size overcurrent device is permitted?

A) The next lower standard size

B) The next higher standard size

C) A device not exceeding 80% of the cablebus rating

D) A device not exceeding 75% of the cablebus rating

Answer: B

NEC Section 240.4(B) permits the next higher overcurrent device rating to be used if less than 800 amperes.

6-54 If a cablebus system is rated at 94 amperes, what is the maximum allowable overcurrent protection device rating?

A) 90 amperes

B) 95 amperes

C) 100 amperes

D) 105 amperes

Answer: C

94 amperes is not a standard rating. Neither is 95 amperes. Therefore, 100 amperes is the next highest standard rating and may be used on the cablebus system. The remaining answer, 105 amperes, is not a standard rating. NEC Section 240.4(B).

6-55 Mobile home disconnecting means shall be located not less than ____ above finished grade or working platform.

A) 8 feet

B) 6 feet

C) 4 feet

D) 2 feet

Answer: D

NEC Section 550.32(F) requires that the bottom of the enclosure be at least 2 feet above grade.

6-56 The branch-circuit conductors to one or more units of a data processing system shall have an ampacity of what percent of the total connected load?

A) 200

B) 125

C) 100

D) 80

Answer: B

NEC Section 645.5(A) requires that branch-circuit conductors supplying one or more units of a data processing system have an ampacity not less than 125 percent of the total connected load.

6-57 Circuit breakers used as switches for high intensity discharge lighting (HID) must be listed and marked:

A) Listing and marking not required

B) HID

C) Intensity rated

D) SWD

Answer: B

NEC Section 240.83(D) now requires circuit breakers used as switches for HID circuits be listed and marked "HID."

6-58 A circuit breaker with a slash rating of 120/240 volts is rated for what voltage to ground?

A) 120 volts to ground

B) 240 volts to ground

C) Any voltage up to 240

D) None of these

Answer: A

NEC Section 240.85 requires the voltage to ground to be the lower of the 2 values on a slash rated circuit breaker.

6-59 The circuit breaker in question 6-58 is permitted to be installed in which of the following systems?

A) Solidly-grounded system

B) Ungrounded system

C) Ungrounded system in RV parks

D) All of these

Answer: A

NEC Section 240.85 states that slash rated circuit breakers are to be used only on solidly-grounded systems.

6-60 Where a power supply enters a recreational vehicle, what is the maximum distance the supply may run inside the vehicle before it must be provided with overcurrent protection?

A) 6 inches

B) 8 inches

C) 10 inches

D) 18 inches

Answer: D

NEC Section 551.30(E) requires an overcurrent device within 18 inches after the supply enters the recreational vehicle.

6-61 Constant voltage generators, except ac generator exciters, must be protected from overloads by:

A) Inherent design

B) Circuit breakers

C) Fuses

D) Any of the above

Answer: D

NEC Section 445.12(A) recognizes all of these answers as acceptable means of protection against overloads.

6-62 What overcurrent requirements are specified for grounded conductors?

A) An overcurrent device must be provided in parallel for every conductor that is intentionally grounded

B) An overcurrent device must be provided in series for every conductor that is intentionally grounded

C) No overcurrent device is permitted in series with any conductor that is intentionally grounded

D) An overcurrent device must be provided both in series and in parallel for every conductor that is intentionally grounded

Answer: C

NEC Section 240.22 requires that no overcurrent device be connected in series with any conductor that is intentionally grounded, unless one of two conditions listed is met.

6-63 One exception to the rule as stated in Question 6-62 is:

A) Where the overcurrent device opens all conductors of the circuit, including the grounded conductor, and is so designed that no pole can operate independently

B) When used on a three-phase, 4-wire circuit

C) Where each pole may be operated independently

D) None of the above

Answer: A

NEC Section 240.22 permits the intentionally grounded conductor to have overcurrent protection if the conditions of answer A are provided.

6-64 Which of the following means of identifying a grounded conductor No. 6 AWG or smaller is not permitted?

A) White outer finish

B) Gray outer finish

C) Three white stripes on other than green insulation

D) Natural gray outer finish

Answer: D

The NEC has removed the term "natural gray" throughout its articles. Conductors must now be distinctly gray if that color is used. NEC Section 200.6(A).

6-65 Where must overcurrent devices in non-power-limited fire alarm circuits be located?

A) At the load on the circuit

B) At the main service panel

C) At the point where the conductor to be protected receives its supply

D) None of the above

Answer: C

NEC Section 760.24 requires the overcurrent protection to be located at the point where the conductor to be protected receives its supply.

6-66 What is the smallest standard plug-fuse rating for use on residential 120-volt branch circuits?

A) 10 amperes

B) 12 amperes

C) 15 amperes

D) 20 amperes

Answer: C

The standard sizes for plug fuses are 15, 20, 25, and 30 amperes. Therefore, 15-ampere fuses are the smallest standard size. NEC Section 240.6(A).

6-67 In general, feeder and branch-circuit conductors must be protected by overcurrent-protective devices connected:

A) Within 2 feet of the point where the conductors receive their supply

B) Within 4 feet of the point where the conductors receive their supply

C) At the immediate point where the conductors receive their supply

D) Within 12 feet of the point where the conductors receive their supply

Answer: C

NEC Section 240.21 requires that overcurrent-protective devices be connected to the branch-circuit conductors at the point the conductors receive their supply; that is, within a reasonable space (a few inches) to allow for either circuit breakers or fuse blocks.

6-68 Branch-circuit overcurrent devices are not permitted in which of the following areas?

A) In residential or motel bathrooms

B) In residential unfinished basements

C) In motel utility rooms

D) In shops containing motor-driven tools

Answer: A

NEC Section 240.24(E) lists residential and motel bathrooms as areas where branch-circuit overcurrent devices shall not be located.

Chapter 7

Utilization Equipment

According to Article 100 of the *NEC*, utilization equipment is equipment that utilizes electric energy for electronic, electromechanical, chemical, heating, lighting, or similar purposes. Therefore, this category includes such items as electric heaters, air conditioners, both large and small appliances, electric motors, lighting fixtures, and the like. Electric motors and motor controllers are covered in Chapter 9, while equipment installed in hazardous locations is covered in Chapter 10.

Utilization equipment differs from other electrical devices in that the former actually uses electric power while the latter are used for control and distribution of electric power. For example, a lighting fixture containing electric lamps is classified as utilization equipment because electric power is required for it to function. A light switch, on the other hand, does not actually use electric power; it merely controls the power to the lamp. Therefore, a manually-operated light switch is a device used to control electricity; it does not utilize electricity to operate. Even lighting contactors that utilize a small amount of electricity to operate their holding coil are not classified as utilization equipment. These devices are classified as controllers.

Large appliances, such as electric ranges, dishwashers, washers, dryers, water heaters, and the like are all classified as utilization equipment. The *NEC* defines the rules for such appliances in terms of their characteristics as well as by the method in which they are connected to their source of electric supply. In many cases, each different type of utilization equipment will have different *NEC* requirements, and the questions appearing on electrician's exams will reflect these differences. For example, the grounding requirements for portable appliances differ from those that are permanently installed. Furthermore, a safety switch or circuit breaker disconnecting a motor-driven appliance rated more than $^1/_8$ horsepower must be in sight of the motor controller to qualify as a disconnecting means: and the switch or circuit breaker must disconnect all ungrounded conductors simultaneously.

There are also separate and distinct *NEC* installation requirements for heating units and air conditioners. The journeyman or master electrician must be familiar with distinctions, and you will find several questions on any electrician's exam that apply to this.

7-1 What is the maximum weight, in pounds, of a lighting fixture when supported only by the screw shell of the lampholder?

A) 2 pounds
B) 4 pounds
C) 6 pounds
D) 8 pounds

Answer: C

NEC Section 410.15(A) limits the weight to 6 pounds.

7-2 What are the maximum dimensions of a lighting fixture that may be supported only by the screw shell of the lampholder?

A) 12 inches
B) 16 inches
C) 18 inches
D) 20 inches

Answer: B

NEC Section 410.15(A) limits these dimensions to 16 inches. That is, any dimension — diameter, length, or width — must not exceed 16 inches if the fixture's only support is the screw shell of the lampholder.

7-3 Flexible cords shall be so connected to devices and to fittings that _____ will not be transmitted to joints or terminals.

A) Tension
B) Shock
C) Heat
D) Voltage

Answer: A

Knotting the cord, winding with tape and fittings designed for the purpose are a few methods commonly used. NEC Section 400.10.

7-4 What is the maximum weight of a lighting fixture that is supported only by an outlet box?

A) 6 pounds
B) 10 pounds
C) 25 pounds
D) 50 pounds

Answer: D

NEC Section 314.27(B) requires a lighting fixture that weighs more than 50 pounds to be supported independently of the outlet box.

7-5 To what portion of a lampholder must the grounded conductor be connected?

A) The base

B) The screw shell

C) The outlet box

D) The pull chain

Answer: B

NEC Section 410.23 requires the grounded conductor, when connected to a screw-shell lampholder, to be connected to the screw shell.

7-6 What is the smallest size fixture wire allowed?

A) No. 14 AWG

B) No. 16 AWG

C) No. 18 AWG

D) No. 20 AWG

Answer: C

NEC Section 410.27(B) states that fixture wires must not be smaller than No. 18 AWG.

7-7 What is the minimum wire size allowed for pendant conductors supplying mogul-base or medium-base screw-shell lampholders?

A) No. 12 AWG

B) No. 14 AWG

C) No. 16 AWG

D) No. 18 AWG

Answer: B

NEC Section 410.27(B) requires a minimum wire size of No. 14 AWG for mogul- or medium-base screw-shell lampholders.

7-8 What is the minimum wire size allowed for pendant conductors supplying intermediate or candelabra-base lampholders?

A) No. 12 AWG

B) No. 14 AWG

C) No. 16 AWG

D) No. 18 AWG

Answer: D

NEC Section 410.27(B) allows conductors not smaller than No. 18 AWG for this application.

7-9 When pendant conductors are longer than a certain length, they must be twisted together. What is this length?

A) 1 foot

B) 2 feet

C) 3 feet

D) 4 feet

Answer: C

NEC Section 410.27(C) requires pendant conductors longer than 3 feet in length to be twisted together where not cabled in a listed assembly.

7-10 Which of the following is an acceptable marking method for the grounding conductor on a grounding-type receptacle?

A) A white-colored round-headed terminal

B) A tag with 1/4-inch high letters

C) An orange dot

D) A green-colored hexagonal headed or shaped terminal screw or nut

Answer: D

NEC Section 406.9(B) permits Answer D as one of the several approved methods of identifying a grounding-type receptacle.

7-11 What type fixture wires are required when the wires are mounted on chains supporting a lighting fixture?

A) Solid conductors only

B) Stranded conductors only

C) Either solid or stranded

D) Compressed aluminum conductors only

Answer: B

NEC Section 410.28(E) requires stranded conductors to be used for wiring on fixture chains and on other movable or flexible parts.

7-12 Flexible cords used to connect cord-connected showcase lighting must be of what type?

A) Light-duty type

B) Extra hard-service type

C) Hard-service type

D) None of the above

Answer: C

NEC Section 410.29(A) requires flexible cords used to connect showcase lighting to be of the hard-service type.

7-13 Normally, lighting fixtures must not be used as a raceway for circuit conductors except when:

A) The fixtures are designed for end-to-end assembly to form a continuous raceway or fixtures connected together by recognized wiring methods shall be permitted to carry through conductors of a two-wire or multiwire branch circuit supplying the fixture

B) The fixtures are listed as a raceway

C) One additional two-wire branch circuit separately supplying one or more of the connected fixtures described in Answer B shall be permitted to be carried through the fixture

D) All of the above

Answer: D

NEC Section 410.32 allows all three conditions in Answers A, B, and C.

7-14 Branch circuit conductors installed within 3 inches of an electric-discharge fixture ballast must have insulation with a temperature rating no lower than:

A) 90°C (194°F)

B) 30°C (86°F)

C) 150°C (302°F)

D) 40°C (104°F)

Answer: A

NEC Section 410.33 requires such conductors to be rated for a temperature not lower than 90°C which includes such types as RHH, XHH, THHN, THHW, FEP, FEPB, SA, and XHHW insulations.

7-15 Electric-discharge lighting fixtures must be plainly marked with their electrical rating. This marking must include the voltage, frequency, and:

A) Wattage

B) Current rating

C) Resistance

D) All of the above

Answer: B

NEC Sections 410.35(A) and (B) require the voltage and frequency to be listed along with the current rating of the unit, including the ballast, transformer, or autotransformer.

7-16 What type of lining must be provided in wiring compartments of all nonmetallic lighting fixtures?

A) Glass

B) Plastic

C) Porcelain

D) Metal

Answer: D

NEC Section 410.37 requires all nonmetallic lighting fixtures not made entirely of noncombustible material to have their wiring compartments lined with metal.

7-17 Which of the following types of lampholders are approved for installation in wet or damp locations?

A) Weatherproof

B) Mogul-base type

C) Intermediate base type with explosionproof fittings

D) Candelabra type

Answer: A

NEC Section 410.49 requires fixtures to be of the weatherproof type when installed in wet or damp locations.

7-18 Isolated ground receptacles shall be identified by:

A) An orange triangle

B) A green dot

C) A permanent seal

D) Standard lugs

Answer: A

NEC Section 406.2(D) requires such receptacles to have an orange triangle located on the face of the receptacle.

7-19 What is the maximum incandescent-lamp wattage that may be used on medium base lampholders?

A) 100 watts

B) 200 watts

C) 300 watts

D) 400 watts

Answer: C

Incandescent lamps for general use on lighting branch circuits that are equipped with medium base lampholders must not exceed 300 watts. NEC Section 410.53.

7-20 What is the minimum spacing required between recessed portions of lighting fixture enclosures and combustible materials, other than at the points of support?

A) $^{1}/_{4}$ inch

B) $^{1}/_{2}$ inch

C) 1 inch

D) 2 inches

Answer: B

NEC Section 410.66 requires at least a $^{1}/_{2}$ inch spacing.

7-21 Incandescent lighting fixtures must be marked to indicate the maximum size lamp permitted. This marking is indicated in:

A) Amperes

B) Total resistance

C) Wattage

D) Watt-hours

Answer: C

Incandescent lighting fixtures are marked to indicate the maximum allowable wattage of lamps. NEC Section 410.70.

7-22 Which of the following is an approved insulating material for lampholders of the screw-shell type?

A) Porcelain

B) Copper

C) Zinc

D) PVC (plastic)

Answer: A

NEC Section 410.72 requires lighting fixtures of the screw-shell type to be constructed of porcelain or other suitable insulating material.

7-23 Auxiliary lighting equipment, not installed as part of the lighting fixture, must be installed in what type of enclosure?

A) Porcelain

B) Glass

C) Metal

D) PVC (plastic)

Answer: C

Auxiliary equipment, including reactors, capacitors, resistors, and similar equipment must be enclosed in an accessible, permanently installed metal cabinet. NEC Section 410.77(A).

7-24 How must a branch circuit supplying a domestic water heater with a capacity of 120 gallons or less be sized?

A) 100% of its nameplate rating

B) 110% of its nameplate rating

C) 125% of its nameplate rating

D) 150% of its nameplate rating

Answer: C

NEC Section 422.13 requires all branch circuits feeding a fixed storage-type water heater having a capacity of 120 gallons or less to have a rating not less than 125% of the nameplate rating of the water heater.

7-25 If a 60 gallon water heater has a nameplate rating of 4500 watts and is fed by a 2-wire, 240-volt circuit, what is the minimum size NM cable conductors that can be used?

A) No. 14 AWG

B) No. 12 AWG

C) No. 10 AWG

D) No. 8 AWG

Answer: C

4500/240 = 18.75 amperes; 18.75 amperes × 1.25 = 23.43 amperes. Since No. 12 AWG NM cable is rated for a maximum load of 20 amperes, the next higher wire size (No. 10 AWG) must be used.

7-26 What is the maximum overcurrent protection that can be provided for the circuit in Question 7-25?

A) 20 amperes

B) 30 amperes

C) 40 amperes

D) 50 amperes

Answer: B

Since No. 10 AWG NM cable is rated for a maximum of 30 amperes, the overcurrent protection device cannot exceed this amperage.

7-27 What is the maximum size infrared heating lamp allowed to be used with medium-base, unswitched porcelain type lampholders?

A) 100 watts

B) 200 watts

C) 300 watts

D) 400 watts

Answer: C

NEC Section 422.48(A) limits the wattage for this situation to 300 watts.

7-28 Wall-mounted ovens and counter-mounted cooking units are permitted to be:

A) Cord- and plug-connected

B) Permanently connected

C) Both A and B

D) Neither A or B

Answer: C

NEC Section 422.16(B)(3) permits these appliances to be either cord- and plug-connected or permanently connected. The NEC has removed the requirement that the cord and plug combination not be used as the required disconnecting means. However, it must be accessible if used as such.

7-29 What is the maximum permitted weight of a ceiling fan mounted directly to an outlet box?

A) 75 pounds

B) 70 pounds

C) 40 pounds

D) 35 pounds

NEC CHANGE

Answer: B

NEC Section 314.27(D) states that a ceiling fan weighing up to 70 pounds may be used provided the box is rated and marked as suitable for this purpose. Outlet boxes designed to support over 35 pounds require markings to include the maximum weight to be supported.

7-30 When the branch-circuit overcurrent device serves as a disconnecting means for a permanently connected appliance, what is the maximum volt-ampere rating allowed for the appliance?

A) 120 volt-amperes

B) 220 volt-amperes

C) 240 volt-amperes

D) 300 volt-amperes

Answer: D

NEC Section 422.31(A) restricts the voltage for such appliances to 300 volt-amperes. If the appliance is rated at more than 300 volt-amperes, the branch-circuit overcurrent protective device may still be used as the disconnect, but it must be within sight of the appliance protected.

7-31 What is the maximum horsepower allowed on permanently connected appliances that use the branch-circuit overcurrent protection means as a disconnect?

A) 1/8 horsepower

B) 1/4 horsepower

C) 3/4 horsepower

D) 1 horsepower

Answer: A

NEC Section 422.31(A) restricts such appliances to 300 volt-amperes or 1/8 horsepower.

7-32 Where the appliance exceeds the maximum ratings specified in Question 7-31, what is one requirement that will still allow the appliance to be disconnected by the branch-circuit overcurrent protection means?

A) The switch or circuit breaker is out of sight of the appliance

B) Double-pole circuit breakers are used

C) Only single-pole circuit breakers are used

D) The switch or circuit breaker is within sight of the appliance

Answer: D

NEC Section 422.31(B) permits a branch-circuit overcurrent protection device to be used as a disconnecting means if the device is within sight of the appliance.

7-33 Name another condition that permits permanently connected appliances to utilize the branch-circuit overcurrent protection device as a means of disconnecting the appliance.

A) The device must be capable of being locked in the open (off) position

B) The device must be a fuse block

C) The device must be enclosed in a metal housing

D) The device must be of the circuit breaker type

Answer: A

NEC Section 422.31(B) permits the branch-circuit overcurrent protection device to serve as a disconnecting means for a permanently connected appliance rated over 300 volt-amperes if the device is capable of being locked in the open position.

7-34 If an external switch or circuit breaker is used for the disconnecting means for a permanently installed appliance containing a motor over 1/8 horsepower, where must the switch or circuit breaker (disconnecting means) be located?

A) On a sheet of plywood firmly secured to a masonry wall

B) On a masonry wall

C) Out of sight of the appliance

D) Within sight of the appliance

Answer: D

NEC Section 422.31(B) requires that the disconnecting means for such appliances be located within sight of the appliance.

7-35 What is the maximum overcurrent protection rating allowed on infrared heating lamps used on commercial or industrial applications?

A) 40 amperes plus 125% of the amperage above 100 amperes

B) 50 amperes

C) 60 amperes

D) 100 amperes plus 125% of the amperage above 100 amperes

Answer: B

NEC Section 422.11(C) restricts the overcurrent protection rating to 50 amperes or less.

7-36 Each electric appliance must be provided with a nameplate giving the identifying name and rating of the appliance in:

A) Volts and watts

B) Volts and amperes

C) Either A or B

D) Neither A or B

Answer: C

NEC Section 422.60 permits either A or B to be used in identifying the electrical characteristics of an electric appliance.

7-37 Under what condition must the frequency be given on the nameplate of an electric appliance?

A) When the appliance is designed for use in foreign countries

B) When the appliance contains resistance heating elements only

C) When the appliance is designed for a specific frequency

D) When the appliance contains heating elements, indicating lights, built-in frequency meter, as well as instrumental controls for operation

Answer: C

The majority of electrical systems in the United States and Canada are designed for 60 Hz. However, the NEC requires the frequency to be indicated on a nameplate if the appliance is designed to be used on a specific frequency. NEC Section 422.60.

7-38 Where electric space heating equipment is supplied by more than one power source, how must the disconnecting means be installed?

A) Installed in separate locations

B) On a plywood panel

C) Grouped and marked

D) On a masonry wall

Answer: C

NEC Section 424.19 requires that the switches be grouped and marked.

7-39 When the branch-circuit overcurrent protection device is used as the disconnecting means for fixed electric space heating with an integral electric motor, what is the maximum horsepower rating the motor can have?

A) $1/8$ horsepower

B) $1/4$ horsepower

C) $1/2$ horsepower

D) 1 horsepower

Answer: A

NEC Section 424.19(B)(1) limits the motor size to $1/8$ horsepower.

7-40 When branch-circuit overcurrent protection is used for motor-driven electric space heating equipment with a motor rated over $1/8$ horsepower, name one requirement for the installation to comply with the *NEC*.

A) The overcurrent device is not capable of being locked open

B) The overcurrent device is within sight of the heater

C) The heater does not exceed 300 volt-amperes

D) The heater does not exceed 4000 watts

Answer: B

NEC Section 424.19(A)(2)(1).

7-41 What is the maximum overcurrent protection allowed for the protection of resistance-type electric space heating equipment?

A) 40 amperes

B) 50 amperes

C) 60 amperes

D) 70 amperes plus an additional 150% if motor driven

Answer: C

NEC Section 424.22(B) limits the maximum overcurrent device to 60 amperes. When higher output is desired, heaters are split and fed with more than one circuit.

7-42 Each unit of fixed electric space heating equipment must be provided with a nameplate giving the identifying name and the normal rating in:

A) Volts and watts

B) Volts and amperes

C) Either A and B

D) Neither A nor B

Answer: C

NEC Section 424.28(A) permits either A or B in marking manufacturers' nameplates for fixed electric space heating.

7-43 What is the maximum distance (interval) that electric heating cable is to be secured during installation?

A) 12 inches

B) 14 inches

C) 15 inches

D) 16 inches

Answer: D

NEC Section 424.41(F) requires ceiling heat cables to be secured at intervals not exceeding 16 inches. This distance may be increased to 6 feet for cables identified for such use.

7-44 By definition, what are duct heaters?

A) Any heater mounted in the air stream of a forced-air system where the air moving unit is not provided as an integral part of the equipment

B) An electric heater that is an integral part of a wall-mounted forced-air heater

C) Any electric heater that is designed for use in wet or damp locations

D) Any heater that is mounted along the baseboard of a room

Answer: A

NEC Section 424.57 defines duct heaters as any heater mounted in the air stream of a forced-air system where the air moving unit is not provided as an integral part of the equipment.

7-45 What is the maximum wattage allowed per linear foot of heating cable embedded in concrete?

A) 10 watts

B) 12½ watts

C) 16½ watts

D) 20 watts

Answer: C

NEC Section 424.44(A) limits the wattage per linear foot for this type of heating cable to 16½ watts.

7-46 What is the minimum spacing between runs of concrete embedded heating cable?

A) 1 inch

B) 2 inches

C) 3 inches

D) 4 inches

Answer: A

NEC Section 424.44(B) requires that the spacing between adjacent runs of cable shall not be less than 1 inch on center.

7-47 On what basis must the branch-circuit conductors and overcurrent protective devices be calculated for an electrode-type boiler?

A) 100% of the total load, not including motors

B) 125% of the total load, including motors

C) 125% of the total load, not including motors

D) 150% of the total load, including all electrical connections

Answer: C

NEC Section 424.82 requires that circuits for electrode-type boilers be calculated on the basis of 125% of the total load, excluding any motors.

7-48 How must the ampacity of branch-circuit conductors and overcurrent protective devices be sized for outdoor electric deicing or snow-melting equipment?

A) 100% of the total load

B) 110% of the total load

C) 120% of the total load

D) 125% of the total load

Answer: D

NEC Section 426.4 requires that conductors and overcurrent protection be sized at 125% of the total load for outdoor deicing and snow-melting equipment.

7-49 On cord- and plug-connecting deicing equipment, the plug itself may be used as the disconnecting means provided:

A) The equipment does not exceed 480 volts

B) The equipment does not exceed 240 volts

C) The equipment does not exceed 150 volts to ground

D) The equipment does not exceed 277 volts to ground

Answer: C

NEC Section 426.50(B) limits the voltage to 150 volts to ground.

7-50 What is the maximum wattage permitted for embedded deicing and snow-melting resistance-type heating units?

A) 100 watts per square foot

B) 120 watts per square foot

C) 150 watts per square foot

D) 200 watts per square foot

Answer: B

NEC Section 426.20(A) limits the wattage to 120 watts per square foot.

7-51 What are the requirements for nonheating leads if they are to be embedded in masonry or asphalt in the same manner as heating cables?

A) They must be under 50 volts

B) The leads must encased in a nonmetallic sheath

C) They must be provided with a grounding sheath or braid

D) None of the above

Answer: C

NEC Section 426.22(A) permits nonheating leads with a grounding sheath or braid to be embedded in masonry or asphalt in the same manner as heating cable without additional physical protection.

7-52 Insulating bushings must be used in asphalt or masonry in a deicing or snow-melting system, where:

A) Leads enter conduit or tubing

B) The voltage exceeds 460 volts

C) Leads change wire size

D) The amperage exceeds 45 amperes

Answer: A

NEC Section 426.22(C) requires the use of bushings at all locations where leads enter a conduit or tubing.

7-53 What is the minimum length of free nonheating lead where it enters a junction box for an embedded heating system?

A) 4 inches
B) 6 inches
C) 8 inches
D) 12 inches

Answer: B

NEC Section 426.22(E) requires that the lead be no less than 6 inches in length where it enters a junction box.

7-54 Each factory-assembled deicing or snow-melting heating unit must be legibly marked with its permanent identification symbol, catalog number, and ratings in volts and watts. What is the maximum distance from the end of the cable that these markings can appear?

A) 1 inch
B) 2 inches
C) 3 inches
D) 4 inches

Answer: C

NEC Section 426.25 requires these markings to be within 3 inches of each end of the nonheating leads.

7-55 Unless provisions are made for expansion and contraction, heating elements and assemblies for fixed outdoor deicing systems shall not be installed where:

A) They can be seen
B) They bridge expansion joints
C) The amperage exceeds 10 amperes
D) They are fed with more than 120 volts

Answer: B

NEC Section 426.20(E) restricts the installation to areas other than over expansion joints, unless provisions are made for expansion and contraction.

7-56 What are two means of securing heating cables for outdoor deicing applications?

A) Rigid conduit straps
B) Frames and spreaders
C) EMT straps
D) NM staples

Answer: B

NEC Section 426.20(D) requires frames or spreaders or other approved means to secure heating cables while the masonry or asphalt finish is applied.

7-57 Which is an approved method for protecting the leads for heating cables as they leave a concrete or asphalt surface?

A) Rigid metal conduit

B) Intermediate metal conduit

C) Rigid nonmetallic conduit

D) All of the above

Answer: D

NEC Section 424.98(E) permits the use of all methods named in Answers A, B, and C.

Chapter 8

Measuring and Testing

Many studies about the nature of electricity begin with the study of the electron and electron theory which states that all matter has electrical charges in various combinations. Actually, the exact nature of electricity cannot be defined, but it may be classified as *static electricity* when the electrons are at rest and *dynamic electricity* when they are in motion. The movement of these electrons is called *current*.

It is more important to know how electricity can be controlled, how to select, install, and maintain electrical systems and equipment, and what to do when problems develop, than it is to know what electricity really is.

In order for current to exist in an electrical circuit, electrons must be in motion. Electrons standing still are no more useful, as far as doing work is concerned, than is water standing still in a garden hose. On the other hand, electricity in motion provides the most effective means yet discovered for carrying energy from one place to another, and for changing one form of energy to another form of energy. To be useful, however, this energy must be controlled or regulated, and knowing the exact quantities of current, voltage, and resistance is one requirement of proper control.

To take accurate readings on electrical measuring instruments for preventive maintenance records and troubleshooting diagnoses, the user should know and apply modern testing techniques and have a good understanding of the characteristics of the basic test instruments. The electrician must also know how to select the correct instrument for the test desired.

This chapter is designed to test your knowledge of testing instruments and how they are used on practical applications. You will also be quizzed on electrical quantities.

8-1 The quantity of electricity used by the consumer is measured by a device called:

A) Ohmmeter

B) Power factor meter

C) Watt-hour meter

D) Ammeter

Answer: C

Electricity is sold by the kilowatt hour and the watt-hour meter is used to register the amount of electricity consumed.

8-2 Which direction do the dials on a kilowatt-hour meter move?

A) Clockwise

B) Counterclockwise

C) Neither answer A or B

D) Both A and B are correct

Answer: D

Look at the illustration in Figure 8-1. Note that the zero mark is at the top of each of the four dials. However, the numbers on the dial to the very right start with the numeral 1 to the right of zero or in a clockwise fashion. The second dial from the right, however, has its numeral 1 to the left of the zero; thus, this dial moves in a counterclockwise direction. Therefore, in most cases, dials on kilowatt-hour meters alternate between clockwise and counterclockwise.

8-3 By what quantity does the dial farthest to the right in Figure 8-1 count kilowatt hours?

A) Singly

B) By tens

C) By hundreds

D) By thousands

Answer: A

For example, if 9 kilowatts are consumed constantly in the circuit to which the meter is connected, the right-hand dial will move three digits in twenty minutes or nine digits in one hour, showing that in one hour 9 kilowatt hours of electric power have been consumed. The second dial from the right counts by tens, the third dial by hundreds, and the left-hand dial by thousands.

Figure 8-1: Typical four-dial kilowatt-hour meter dials

8-4 What is the reading of the meter in Figure 8-1?

A) 1, 2, 3, 9 or 1239 kilowatt hours

B) 2, 1, 3, 9 or 2139 kilowatt hours

C) 2, 2, 3, 9 or 2239 kilowatt hours

D) 1, 1, 2, 8 or 1128 kilowatt hours

Answer: D

When reading kilowatt-hour meters, you read the number that the dial has just passed. In the left-hand dial, the hand has passed the numeral 1, but has not yet reached the numeral 2. Therefore, the reading is 1. The same is true for the second dial from the left, and so on.

8-5 Some kilowatt-hour meters have five or more dials like the one shown in Figure 8-2. In a five-dial meter, what amount does the very left-hand dial count?

A) Hundreds

B) Tens

C) Thousands

D) Ten thousands

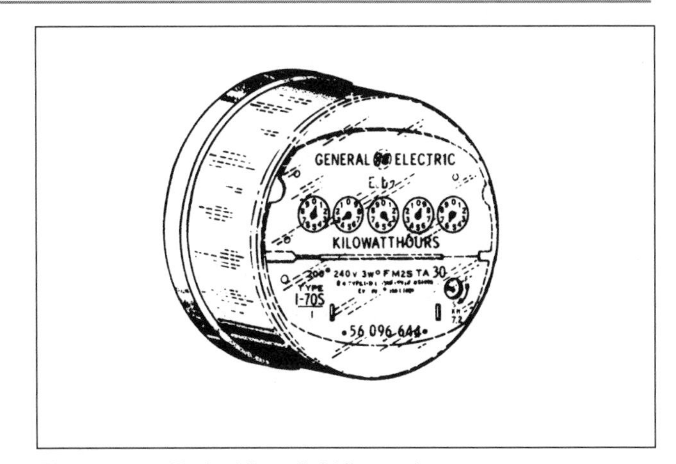

Figure 8-2: Typical five-dial kilowatt-hour meter

Answer: D

Reading from right to left, a five-dial kilowatt-hour meter reads singly, tens, hundreds, thousands, and ten thousands. Therefore, the dial farthest to the left reads ten thousands in a five-dial meter.

Figure 8-3: An example of a five-dial meter reading

8-6 What is the reading of the kilowatt-hour meter in Figure 8-3?

A) 22,179

B) 33,179

C) 34,179

D) 35,179

Answer: A

Reading from left to right, the first dial has passed the numeral 2, but has not yet reached 3. Therefore, this dial reads 2; the same is true for the next dial. The third dial from the left has passed the numeral 1, but has not yet reached 2; the reading is 1. The next dial has passed the numeral 7, but has not yet reached 8; thus, the reading is 7. The next dial (the one farthest to the right) has passed the numeral 9. Therefore, the meter reading is 22,179 kilowatt-hours.

8-7 Electricity is sold by the:

A) Ampere hour

B) Kilowatt hour

C) Volt hour

D) None of the above

Answer: B

See explanation for Question 8-1.

8-8 If a three-phase circuit is not balanced, how many single-phase wattmeters are necessary to measure the power in the circuit?

A) 1
B) 2
C) 3
D) 4

Answer: C

In an unbalanced three-phase circuit, it becomes necessary to use three single-phase wattmeters — one in each ungrounded conductor — to obtain the total power in the circuit.

8-9 How many wattmeters are necessary to measure the power in a balanced three-phase, Y-connected circuit?

A) 1
B) 2
C) 3
D) 4

Answer: A

A balanced Y-connected, three-phase load is equivalent to three equal single-phase circuits that have a common neutral conductor. The power in such a circuit can be measured by using one single-phase wattmeter connected between one ungrounded phase and the grounded neutral. The total power of the circuit may be determined by multiplying the reading on the wattmeter by three.

8-10 If a completely balanced Y-connected, three-phase load shows a power reading of 7 kilowatt hours using a single wattmeter, what is the total power of the circuit?

A) 20 kilowatt hours
B) 21 kilowatt hours
C) 22 kilowatt hours
D) 23 kilowatt hours

Answer: B

$3 \times 7 = 21$ kilowatt hours. Also see explanation for Question 8-9.

8-11 What is the name of a direct-indicating instrument for measuring resistance?

A) Voltmeter

B) Wattmeter

C) Ohmmeter

D) Ammeter

Answer: C

The essential parts of an ohmmeter are a source of voltage, a resistor, and a voltmeter. The voltage is normally suppied by a conventional battery and is usually connected to a variable rheostat so that the battery voltage can be adjusted to a constant predetermined value. The resistor is connected in series with the voltmeter and the lead terminals that are connected to the unknown resistance to be measured. When the reading is taken, no computation is necessary; the instrument indicates resistance directly on a scale calibrated in ohms.

8-12 A measuring instrument consisting of an enclosure containing resistance coils with convenient plugs or switches for placing coils of various resistance in and out of a balancing circuit is called a:

A) Volt-ohm-ammeter

B) Voltmeter

C) Current transformer

D) Wheatstone bridge

Answer: D

The Wheatstone bridge is a convenient device for measuring the resistance of electrical circuits or components by comparison with a standard resistance of known values.

8-13 The power factor of a system or circuit is:

A) The ratio of true power to apparent power

B) The ratio of impedance to reactance

C) The relationship of volts and ohms

D) None of the above

Answer: A

The power factor of a circuit can be obtained by taking simultaneous readings with an ammeter, a voltmeter, and a wattmeter and then dividing the watts by the product of the voltage and current; that is:

$$PF\ (\text{single-phase circuit}) = \frac{watts}{volts \times current}$$

Apparent power is the theoretical power without taking any circuit losses into consideration. True power is the actual power consumed by the system. The ratio of the two gives the power factor of the system.

8-14 An instrument used to show the difference in phase and frequency between the voltages of two alternators is called:

A) Tachometer

B) Synchroscope

C) Light meter

D) None of the above

Answer: B

When two alternators are about to be connected in parallel, the voltages of the two must be approximately the same; their voltages must be exactly in phase; and their frequencies must be approximately the same. If these differences are too great, the alternators are likely to pull entirely out of phase, thus causing a complete shutdown. The synchroscope indicates what adjustments are necessary to meet the above conditions.

8-15 An instrument used to indicate or record the speed of a machine in revolutions per minute (rpm) is called:

A) Frequency meter

B) Footcandle meter

C) Tachometer

D) Synchroscope

Answer: C

Several designs of tachometers are available: centrifugal, eddy current, surface speed, vibrating reed, and high-intensity stroboscope or photo types. Some tachometers are connected to a machine by means of belts or gears while others are hand held. All have various scales from 0 to 50,000 rpm.

8-16 Modern ac power systems must be closely regulated to a constant frequency — within 1%. What is the name of the very sensitive instrument used to measure these small variations?

A) Frequency meter

B) Footcandle meter

C) Tachometer

D) Synchroscope

Answer: A

Most frequency meters are either of the ratio-meter design or the vibrating-reed design similar to the vibrating-reed tachometers. Either type may be used to measure the frequency of any ac electrical circuit.

8-17 Which of the following instruments are used to measure temperatures?

A) Resistance thermometer

B) Radiation-pyrometer

C) Thermocouple thermometer

D) All of the above

Answer: D

Answers: A, B, and C are all used to measure temperatures. Resistance thermometers are used to measure temperatures up to about 1500°F. The thermocouple method is used for temperatures up to about 3000°F. The radiation-pyrometer and optical-pyrometer are used for measuring temperatures above 3000°F.

8-18 When taking low-current readings with a clamp-on ammeter, what device is normally used to help get a more accurate reading on the ammeter scale?

A) Frequency meter

B) Current multiplier

C) Phase-sequence indicator

D) Current transformer

Answer: B

A current multiplier allows current measurement on low-current equipment since the load current is multiplied by either 2, 5, or 10 times; that is, if the meter scale shows a reading of 62 amps and the 10 × multiplier is used, the actual load current would be 62/10 or 6.2 amperes.

8-19 What is another way to multiply the current reading on an ammeter's scale when a current multiplier is not available?

A) Wrap several loops of a single circuit conductor around the hook of the clamp-on ammeter

B) Try to calculate the current by some other means

C) Use a magnifying glass to read the meter's scale

D) None of the above

Answer: A

Wrapping a single conductor around the hook of the ammeter will multiply the current as shown in Figure 8-4.

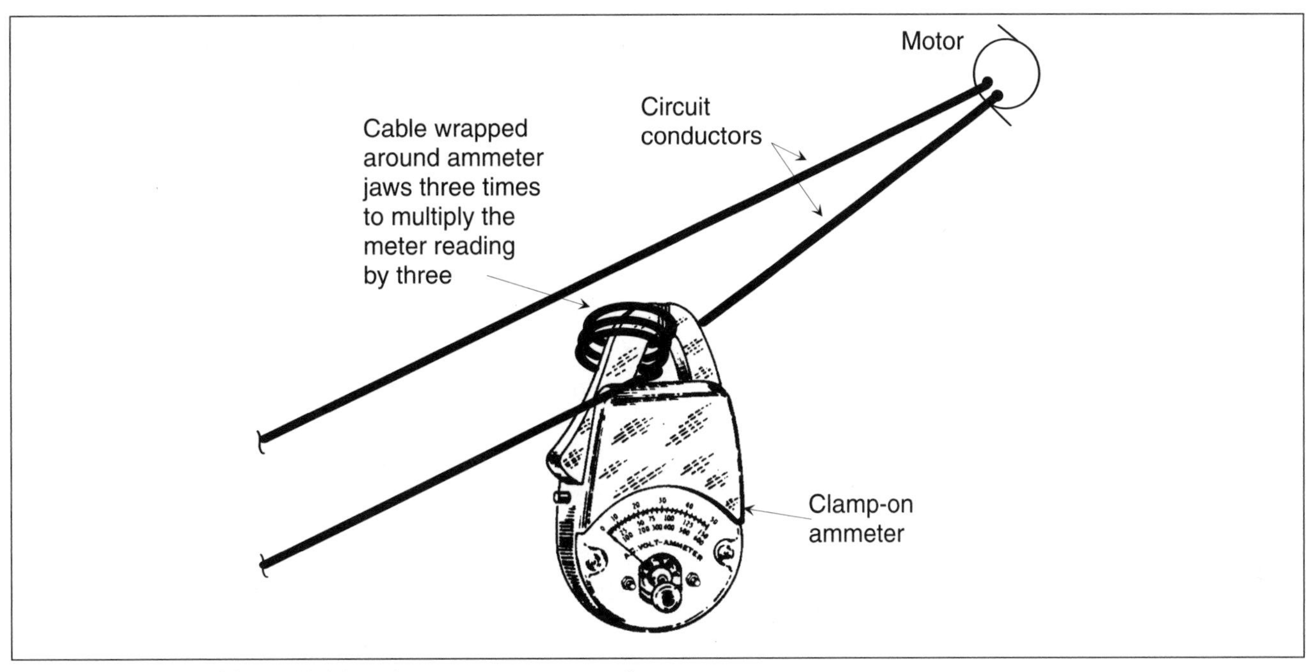

Figure 8-4: Method for multiplying clamp-on ammeter reading

8-20 The ammeter in Figure 8-4 shows three turns of wire wrapped around the ammeter's hook. If the actual current drawn by the motor is 5 amperes, what will the meter reading be?

A) 5 amperes

B) 10 amperes

C) 15 amperes

D) 20 amperes

Answer: C

The current is multiplied by the number of coils wrapped around the hook. Since this illustration shows three loops, the current (5 amperes) is multiplied by 3 giving a meter reading of 15 amperes.

8-21 How should a voltmeter be connected in a circuit to take a voltage reading under normal conditions?

A) In series

B) In parallel

C) In series/parallel

D) None of the above

Answer: B

A voltmeter is always connected across the load or in parallel to obtain a voltage reading. Ammeters are connected in series with the load.

8-22 If a voltage test is made at the main disconnect on a 120/240-volt, three-phase, 4-wire delta system, what should the voltage be between any two ungrounded conductors?

A) 120 volts

B) 208 volts

C) 240 volts

D) 250 volts

Answer: C

Correct voltage readings between either phases A and B, A and C, or B and C will be approximately 240 volts on a delta-connected, 4-wire system. Each ungrounded conductor in an electrical system is identified by letters: A and B phases in a single-phase system; A, B, and C phases in a three-phase system.

8-23 What will the voltages be between the grounded conductor and any ungrounded conductor in the electric service in Question 8-22?

A) 120 volts

B) 208 volts

C) 240 volts

D) 120 volts on two phases

Answer: D

Voltage readings will be 120 volts except on the "high leg" phase which is usually between 180 and 190 volts.

8-24 If the service in Question 8-22 is supplied by a utility company, and the voltage reading at the main service equipment is below 230 volts, who is usually responsible for correcting the low-voltage problem?

A) The building's owners

B) The utility company

C) The building's tenants

D) The NFPA

Answer: B

Since the reading is taken at the main service equipment before the electric service encounters any load or long feeder runs, the correction of the problem lies with the utility company. To correct, the transformer taps are normally changed to increase the voltage to a normal operating level of between 230 and 240 volts.

8-25 If a voltage reading is taken between any two ungrounded conductors on a 120/208-volt, three-phase, 4-wire Y-Y-connected electric service, what should the normal voltage reading be?

A) 120 volts

B) 208 volts

C) 240 volts

D) 250 volts

Answer: B

The voltage between any two phase conductors is calculated by the sum of all three phase conductors divided by the square root of 3. Thus, 120 + 120 + 120 = 360 divided by 1.73 = 208.092 volts. The voltage may also be calculated by multiplying the square root of three by any one phase conductor; that is, 120 × 1.73 = 207.6 volts. In either case, the resulting voltage is approximately 208 volts.

8-26 In the electric service in Question 8-25, what is the voltage between the grounding conductor and any phase conductor?

A) 120 volts

B) 208 volts

C) 240 volts

D) 120 volts on two of the phases and approximately 190 volts on the remaining phase

Answer: A

On a three-phase, 120/208-volt, Y-connected electric service, the voltage between the grounding conductor and any phase or ungrounded conductor is the same on all three phases.

8-27 If a voltage reading is taken at the main service disconnect for the electric service in Question 8-25 and found to be approximately 208 volts between any two ungrounded conductors, but the same voltage test shows only 190 volts at the last outlet on the circuit, where does the fault lie?

A) With the utility company

B) With the building's interior wiring

C) With the power company's transformer

D) None of the above

Answer: B

Since correct voltage was measured at the main service disconnect, the voltage loss occurred within the building's wiring system — probably due to long feeder or circuit runs. The problem can be corrected by increasing the wire size of the feeders and/or branch circuits. Buck-and-boost transformers may also be connected in the circuit to maintain the correct voltage.

8-28 If voltage readings are taken on a three-phase, 4-wire, 240-volt, delta-connected electric service and two of the phases show a reading of 230 volts between phase and ground and the third shows a reading of only 50 volts between phase and ground, what is the probable cause of this problem?

A) Load too great

B) Incorrect wire size

C) Incorrect transformer settings

D) A partial ground or ground fault

Answer: D

The phase with the lowest voltage reading has a partial ground or ground fault.

8-29 When cartridge fuses are tested with a voltmeter, they should be tested:

A) Phase-to-phase on the line side

B) Diagonally with one lead on the line side of one fuse and the load side of an adjacent fuse

C) Phase-to-phase on the load side

D) None of the above

Answer: B

The fuses may be checked by testing across diagonally from the line to the load side as shown in Figure 8-5.

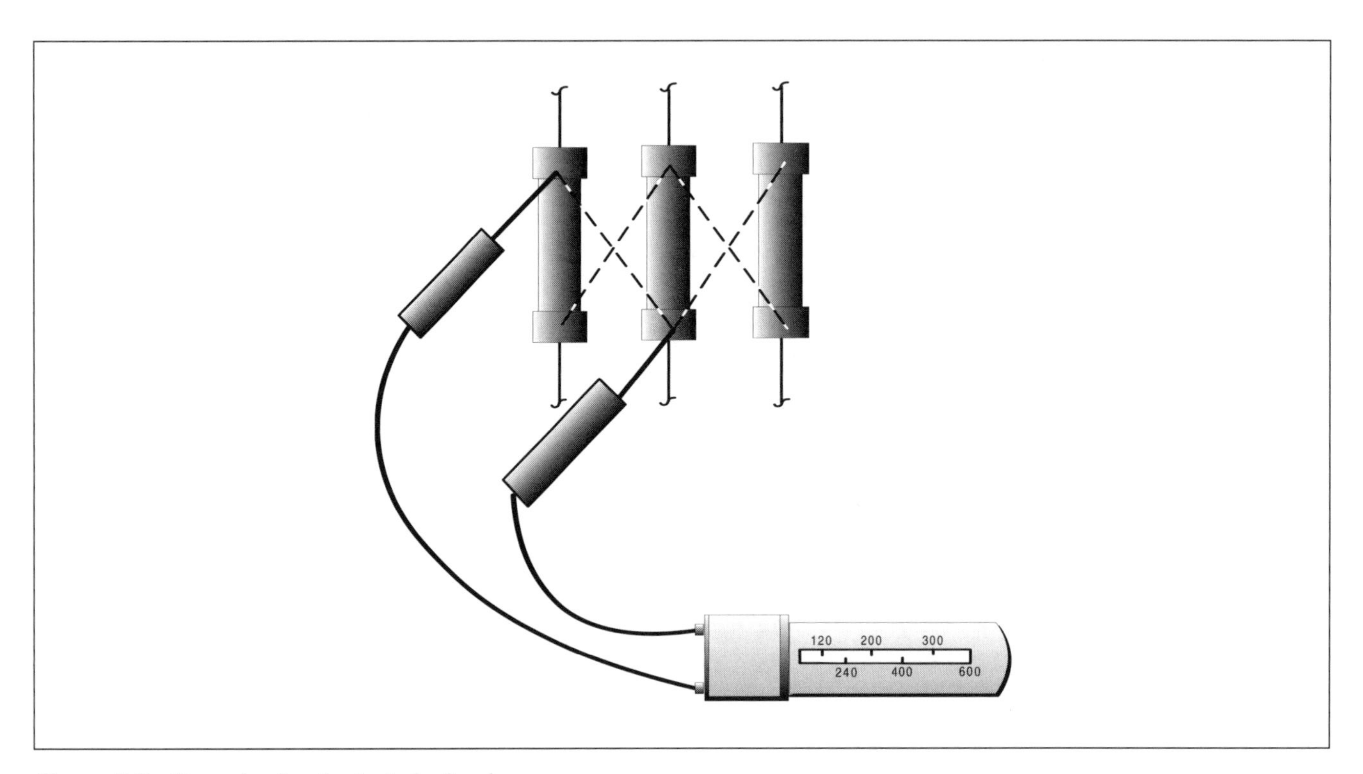

Figure 8-5: Correct voltmeter tests for line fuses

8-30 In a dc circuit, the ratio of watts to volt-amperes is always:

A) Greater than one

B) Less than one

C) Difficult to determine

D) At unity

Answer: D

On a dc circuit, voltage and current happen simultaneously; thus, the power factor is said to be at unity.

8-31 The first step in solving a low-voltage problem on a circuit is to leave the voltmeter leads connected across the line, and then begin:

A) Connecting additional loads to the circuit

B) Disconnecting loads to the circuit

C) Both A and B

D) Neither A nor B

Answer: B

The voltmeter leads are connected to the circuit and are observed after disconnecting each load from the circuit. When the voltmeter shows a normal reading, the load just disconnected probably has a partial ground or ground fault.

8-32 A dc voltmeter may be used to measure a low resistance in a de-energized ac circuit when the voltmeter is used in conjunction with:

A) A megger

B) An ammeter

C) A known resistance

D) A current transformer

Answer: C

To measure an unknown resistance in a circuit with a dc voltmeter, connect another resistor of known resistance in series with the unknown resistor. Then connect a 6-volt battery to the circuit; the same current will pass through both resistors. The voltage drop across each can then be determined and the unknown resistance may be found as explained in Question 8-30.

8-33 A tachometer's output is calibrated in:

A) fps

B) foot pounds

C) cfm

D) rpm

Answer: D

A tachometer measures the revolutions per minute of a revolving object or machine. Therefore, a tachometer's output is calibrated in rpm.

8-34 Which of the following is a good indication that an open circuit exists?

A) A high voltage reading

B) No voltage reading

C) A low voltage reading

D) None of the above

Answer: B

If a circuit is open (not completed) there will be no voltage reading when tests are made.

8-35 One visual test that may be made on a clear, incandescent lamp to see if it is good or not is to:

A) Look for a broken element

B) Look for a corroded base

C) Look for discolored glass

D) None of the above

Answer: A

A broken element or filament in an incandescent lamp is normally visible. A broken element means that the circuit is not completed and the lamp will not light.

8-36 One way to test a frosted incandescent lamp is with a testing instrument, and when no means of electric current is available:

A) With an ohmmeter

B) With a wattmeter

C) With a power factor meter

D) All of the above

Answer: A

The test leads of an ohmmeter may be connected to an incandescent lamp — one lead touching the positive terminal at the base of the lamp and the other touching the shell or threaded portion of the lamp. If the ohmmeter shows resistance, the element is intact and the lamp should be good.

8-37 When testing a load center with the main disconnect open (panel de-energized) using an ohmmeter, a blown or faulty fuse would give a reading of:

A) 0 ohms

B) 17 ohms

C) Infinity

D) 70 ohms

Answer: C

A blown fuse would interrupt the circuit, making it open. Therefore, the ohmmeter would read an infinite resistance.

8-38 Given the same situation as described in Question 8-37, if the ohmmeter showed any resistance at all on the scale, the conclusion would be:

A) The fuse is good

B) The fuse is partially good

C) The fuse is bad

D) The fuse is partially bad

Answer: A

Any resistance reading at all would indicate a complete circuit, meaning that the fuse is good.

8-39 The basic definition of a short circuit is:

A) An undesired current path that does not bypass the load

B) An undesired current path that allows current to bypass the load

C) A desired current path that does not bypass the load

D) A desired current path that bypasses the load

Answer: B

A short circuit or ground fault completes the circuit before reaching the load. In most cases, the overcurrent protective device will open or blow when a short circuit exists on the circuit.

8-40 When testing conductor insulation with a megger, it is desirable to have the instrument's voltage:

A) Slightly lower than the peak value of the rated ac voltage

B) The same as the rated ac voltage

C) Slighly higher than the peak value of the rated ac voltage

D) No more than 80% of the ac voltage

Answer: C

The megger's dc voltage should be slightly higher than the peak value of the rated ac voltage. In doing so, any weaknesses will be revealed to a higher percentage than if the same voltage was used.

8-41 Name the three types of instruments suitable for making insulation tests at two or more voltages.

A) Ammeter, ohmmeter, voltmeter

B) Power-factor meter, tachometer, and ohmmeter

C) Hand-driven, motor-driven, and rectifying meggers

D) Watt-hour meter, power-factor meter, and tachometer

Answer: C

The types in Answer C all are capable of producing more than one voltage level.

8-42 Most meggers provide a range in voltage from:

A) 0 to 50 volts

B) 500 to 1000 volts

C) 0 to 500 volts

D) 500 to 5000 volts

Answer: D

The specifications of most meggers or insulation testers have a voltage range of from 500 to 5000 dc volts.

8-43 In general, what are the three basic electrical faults?

A) Ground fault, partial ground fault, and open circuit

B) Short circuit, open circuit, and change in electrical value

C) Open circuit, incomplete circuit, and ground fault

D) Short circuit, ground fault, and partial ground fault

Answer: B

An open circuit is the most common; a short circuit or ground fault is the next most common problem, and a change in electrical value such as low voltage or higher resistance also frequently causes problems.

8-44 The nature of a problem on a circuit protected by a plug fuse can often be determined by the appearance of the fuse:

A) Base

B) Enclosure

C) Window

D) Terminal conductor

Answer: C

The appearance of the fuse window will frequently give the electrician a clue as to the nature of the problem. See Questions 8-45 and 8-46.

8-45 If a plug fuse window appears clear and the element is merely broken, the probable cause of the fuse blowing is:

A) A short circuit

B) A change in electrical value

C) An overload

D) None of the above

Answer: C

The clear fuse window shows no sign of a violent "explosion" or breakage of the fuse link. Therefore, the most probable cause is an overload on the circuit.

8-46 If a plug-fuse window is blackened or discolored, the most probable cause of the fuse blowing is:

A) A short circuit

B) A change in electrical value

C) An overload

D) None of the above

Answer: A

The blackened or discolored fuse window indicates a sudden and violent eruption of the fuse element. Therefore, a short circuit or ground fault is the most probable cause of the fuse blowing. The circuit should be tested and the fault corrected before replacing the blown fuse.

8-47 One way to test electrical measuring instruments for accuracy is to:

A) Compare the instrument under test with another similar instrument of known accuracy

B) Test the instrument on an electric circuit

C) Compare the instrument under test with another similar instrument; if they both show about the same reading, both are correct

D) None of the above

Answer: A

Comparing the readings with another similar instrument of known accuracy is the simplest way to test a measuring instrument for accuracy.

8-48 To test an ammeter for accuracy, another ammeter of known accuracy can be connected in the circuit with the ammeter under test. How should these ammeters be connected?

A) In parallel

B) One in series, the other in parallel

C) In series with each other

D) None of the above

Answer: C

They must be connected in series.

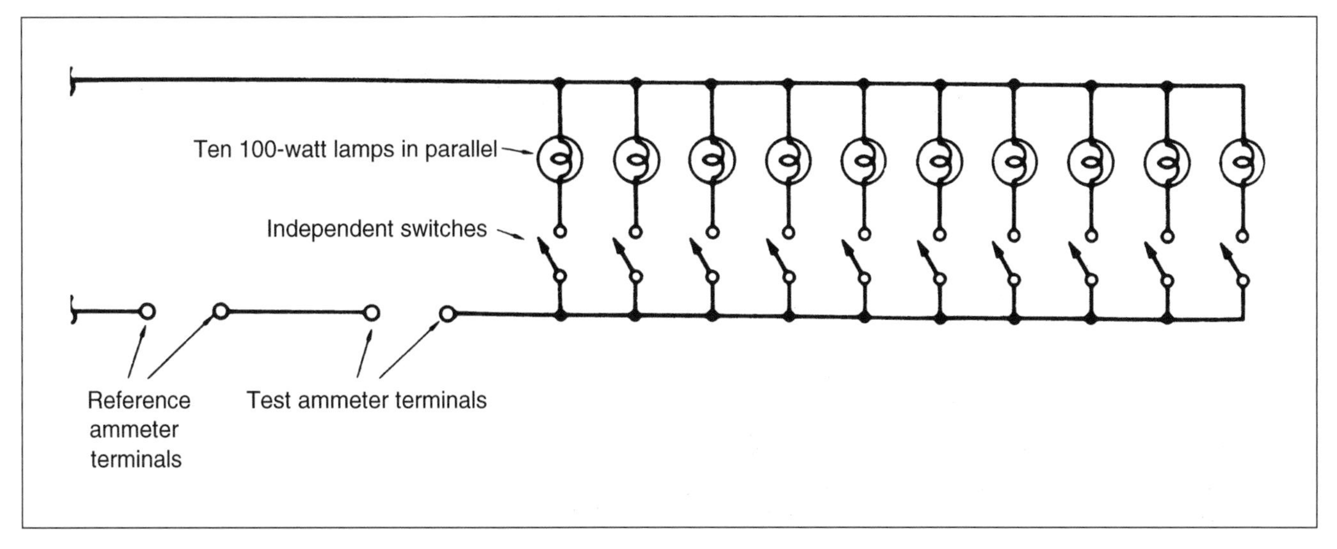

Figure 8-6: Test circuit for calibrating ac and dc ammeters by comparison method

8-49 One type of test circuit is shown in Figure 8-6. If the ten 100-watt lamps are connected to a 118-volt source, what should the reading of both ammeters be?

A) 6.47 amperes

B) 8.47 amperes

C) 10.47 amperes

D) 12.47 amperes

Answer: B

Amperes = number of lamps in circuit × wattage of each lamp/actual voltage = (10 × 100)/118 = 8.47.

8-50 What is the purpose of the independent switches in Figure 8-6?

A) To allow each lamp to be cut off to change the bulb

B) To prevent electrical shock

C) To vary the load on the circuit so the ammeters will show different readings

D) None of the above

Answer: C

To obtain a different ammeter reading, all lamps are first turned off. Then one lamp can be turned on. The ammeter reading should be approximately .84 amperes. Then another lamp is turned on and the reading should be approximately 1.68 amperes and so on. By being able to vary the load in this manner, the ammeter can be tested at different ranges on the scale.

Chapter 9

Electric Motors and Motor Controls

Induction motors get their name from the fact that they utilize the principle of electromagnetic induction. An induction motor has a stationary part, or stator, with windings connected to the ac supply, and a rotation part, or rotor, which contains coils or bars. There is no electrical connection between the stator and rotor. The magnetic field produced in the stator windings induces a voltage in the rotor coils or bars.

Since the stator windings act in the same way as the primary winding of a transformer, the stator of an induction motor is sometimes called the *primary*. Similarly, the rotor is called the *secondary* because it carries the induced voltage in the same way as the secondary of a transformer.

Induction motors are made in sizes from fractional horsepower to several hundred horsepower and there are few electrical systems that do not use electric motors in some capacity — from residential air conditioners and appliances to commercial HVAC equipment, elevators, and huge motors for industrial applications. Yes, a motor even starts your car in the mornings, or mixes your drinks in an electric blender.

For a motor to be of any practical use, it must be controlled — if only to start and stop it. Therefore, all motors utilize some type of controller. These controllers cover a wide range of types and sizes, from a simple toggle switch to a complex system consisting of such components as relays, timers, switches, pushbuttons, and the like. The common function, however, is the same in every case; that is, to control some operation of an electric motor.

The *National Electrical Code* has certain rules and requirements for motor installations, and questions on electrician's exams normally use these *NEC* rules as a basis for the test questions. This chapter covers many of the questions that have appeared on electrical examinations in the past; many of them will more than likely appear in the future.

9-1 A motor controller that also serves as a disconnecting means:

A) Shall open all ungrounded conductors to the motor simultaneously

B) Shall not be required to open all conductors to the motor

C) Shall open only grounded conductors

D) Shall open only one current-carrying conductor

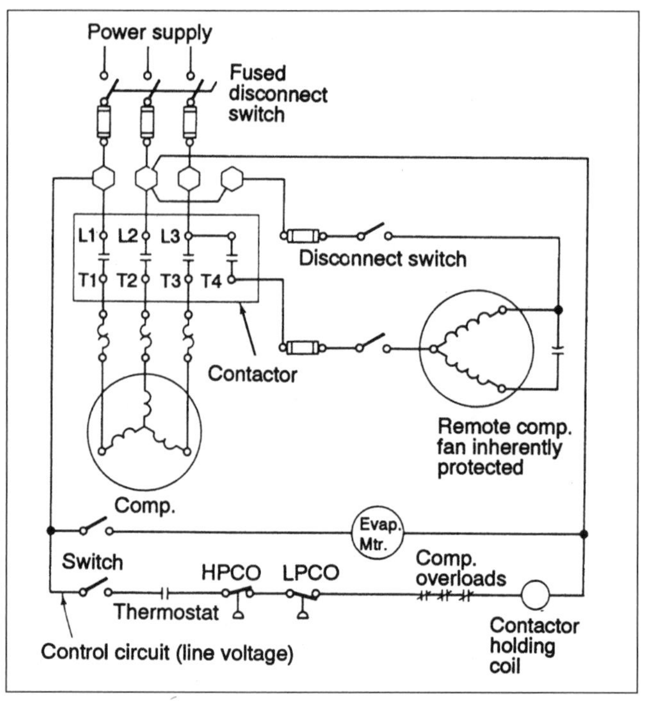

Figure 9-1: All ungrounded conductors must be opened when a motor controller also serves as a disconnect

Answer: A

All ungrounded conductors must be disconnected from the motor simultaneously. NEC Section 430.111(A). See Figure 9-1.

9-2 The branch-circuit overcurrent device may serve as the controller for stationary motors rated less than:

A) 20 horsepower

B) 1/8 horsepower

C) 15 horsepower

D) 5 horsepower

Answer: B

Small motors (1/8 horsepower or less) may use the branch-circuit fuse or circuit breaker as a means of disconnecting and de-energizing the motor. This motor must be the type normally left running (such as a clock motor) and constructed so as not to be damaged by overload. NEC Section 430.81(A).

9-3 A disconnecting means for a 2300-volt motor must be capable of being locked in the:

A) Closed position

B) Open and closed positions

C) Open position

D) ON position

Answer: C

All motors must be connected to a disconnecting means. However, motors over 600 volts must also be provided with a means of locking the disconnect open so that it cannot be inadvertently closed when someone is working on the motor. NEC 430.227.

9-4 Each continuous-duty motor rated at more than what horsepower must be protected against overload by an approved method?

A) 1/8 horsepower

B) 1/2 horsepower

C) 1/4 horsepower

D) 1 horsepower

Answer: D

Smaller motors (under 1 horsepower) such as those used to power fans, drill presses, etc. do not require overload protection. However, the larger motors (1 horsepower or over) that are used to drive heavy machinery do require "heaters" or other types of approved overload protection. NEC Section 430.32(A).

9-5 A disconnecting means that must be visible from a motor location must not be farther than:

A) 25 feet

B) 75 feet

C) 50 feet

D) 80 feet

Answer: C

Where the NEC specifies that one item or piece of equipment shall be "in sight from," "within sight from," or "within sight," etc. of another item or piece of equipment, they must not be more than 50 feet apart. NEC Article 100 — Definitions. See Figure 9-2.

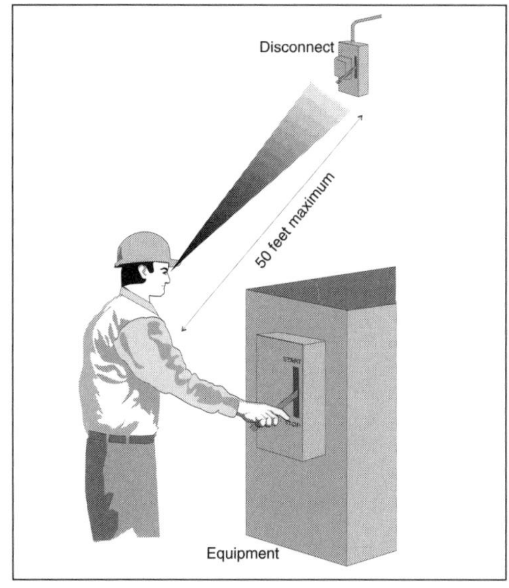

Figure 9-2: *NEC* definition: within sight from

9-6 A 20-ampere ac snap switch may disconnect a 2-horsepower motor with a maximum full-load current of:

A) 16 amperes

B) 10 amperes

C) 20 amperes

D) 18 amperes

Answer: A

A snap switch used to disconnect a motor must be derated 80%. Therefore, 20 amperes × .80 = 16 amperes. NEC 430.109(C)(2).

9-7 Live parts of motors must be guarded if they operate at over:

A) 150 volts
B) 50 volts
C) 100 volts
D) 75 volts

Answer: B

NEC Section 430.232 requires exposed live parts of motors and controllers operating at 50 volts or more between terminals to be guarded against accidental contact.

9-8 The smallest size conduit allowed to enclose motor terminals from a junction box is:

A) 1/2 inch
B) 1 inch
C) 1 1/2 inch
D) 3/8 inch

Answer: D

A junction box to house motor terminals shall be run, in general, in flexible or rigid metal raceway and must not be smaller than 3/8-inch electrical trade size. NEC Section 430.245(B).

9-9 Motor controller enclosures must be grounded:

A) At all voltages
B) If above 600 volts
C) If above 50 volts
D) If 240 volts or more

Answer: A

NEC Section 430.244 states that controller enclosures shall be grounded regardless of voltage and must have a means for attaching an equipment grounding conductor as per NEC Section 250.8.

9-10 A branch circuit is supplying a single hermetic refrigerant motor-compressor for an air conditioning unit in a single-family dwelling. The hermetic refrigerant motor-compressor rated-load current is 18 amperes. If a 30-ampere fuse will not carry the motor compressor's starting current, the maximum rating of the branch-circuit overcurrent protective device may be increased to:

A) 35 amperes
B) 40 amperes
C) 45 amperes
D) 50 amperes

Answer: B

The protective device can be increased also, but shall not exceed 225% of the motor rated-load current or branch-circuit selection current, whichever is greater. NEC 440.22(A).

9-11 The minimum size liquidtight flexible metal conduit for general use is:

A) 3/8 inch

B) 3/4 inch

C) 1/2 inch

D) 1 1/2 inch

Answer: C

Liquidtight flexible metal conduit smaller than 1/2-inch electrical trade size must not be used (with exception, as covered in NEC Section 350.20(A)).

9-12 Which of the following statements about liquidtight flexible metal conduit is/are correct?
I. Liquidtight flexible metal conduit shall not be used where subject to physical damage.
II. Liquidtight flexible metal conduit shall be used only with listed fittings.

A) I only

B) II only

C) Both I and II

D) Neither I nor II

Answer: C

NEC Articles 350.12 and 350.6 verify that both statements are true.

9-13 The conductors of a feeder supplying two (2) continuous duty motors, one rated 7 1/2 horsepower, 240 volts dc, and one rated 10 horsepower, 240 volts dc shall have a current rating not less than:

A) 56 amperes

B) 74 amperes

C) 77 amperes

D) 84 amperes

Answer: C

NEC Table 430.247 lists the 7 1/2 horsepower motor at 29 amps; the 10 horsepower motor at 38 amps. Therefore, 29 + 38 = 67 amps plus 25% of the highest rated motor in the group (9.5 amps) for a total of 76.5 amps which rounds off to 77 amps.

9-14 A motor disconnecting means must:

A) Disconnect only the motor

B) Disconnect only the controller

C) Disconnect both the motor and controller

D) Disconnect the motor

Answer: C

A motor disconnecting means should be installed on the line side of the controller so that both the controller and motor are disconnected simultaneously. NEC Article 430.74.

9-15 If a conductor is covered, it is:

A) Encased with an insulation recognized by the Code

B) Encased with an outer covering not recognized as electrical insulation

C) Effectively isolated from ground faults

D) Covered with a material of a composition and thickness recognized by UL

Answer: B

A covered conductor is a conductor encased within a material of composition or thickness that is not recognized by Code as electrical insulation. NEC Article 100 — Definitions.

9-16 Open motors with commutators shall be located so sparks cannot reach adjacent combustible material, but this:

A) Is only required for over 600 volts

B) Does not prohibit these motors from a Class I location

C) Shall not prohibit these motors on wooden floors

D) None of these

Answer: C

Installation of such motors must be located away from combustible material. However, the code makes exception for installations on wooden floors or supports. NEC 430.14(B) Exception.

9-17 The branch-circuit conductors supplying a motor that draws 40 amperes and operates for three hours or more must have an ampacity not less than:

A) 40 amperes

B) 50 amperes

C) 60 amperes

D) 70 amperes

Answer: B

Branch-circuit conductors supplying a single motor at continuous duty (three hours or more) must have an ampacity not less than 125% of the motor full-load current rating. Therefore, 40 × 1.25 = 50 amperes. NEC 430.22(A).

9-18 If no other information is given, a continuous-duty motor that draws 100 amperes must be protected by an overload device rated at not more than:

A) 125 amperes

B) 150 amperes

C) 140 amperes

D) 115 amperes

Answer: D

NEC Section 430.32(A)(1) states that each continuous-duty motor circuit must be protected against overload by a separate overload device. This device must be selected to trip at no more than 115% (if no other information about the motor is known).

9-19 The branch-circuit overcurrent device may serve as the disconnecting means for motor-driven appliances rated not over:

A) 1/2 horsepower

B) 1 horsepower

C) 3/4 horsepower

D) 1/8 horsepower

Answer: D

For permanently connected appliances rated at not over 300 volt-amperes (watts) or 1/8 horsepower, the branch-circuit overcurrent device may be used to serve as a disconnecting means. NEC Section 422.31(A).

9-20 A power factor correction capacitor connected on the load side of the overload device of a motor reduces the line current by 20%. If the full-load current of the motor is 100 amperes, the setting of an overload device in the circuit is based on a current of:

A) 100 amperes
B) 125 amperes
C) 80 amperes
D) 150 amperes

Answer: C

Overload device setting = full-load current × percent factor; 100 × .80 = 80 amperes.

9-21 Conductors supplying two or more motors must have an ampacity equal to the sum of the full-load current rating of all the motors plus what percent of the highest rated motor in the group?

A) 25%
B) 50%
C) 75%
D) 100%

Answer: A

NEC Section 430.24 requires an addition of 25% of the highest rated motor in the group.

9-22 What may the locked-rotor current be assumed to be for small motor-compressors not having the locked-rotor current marked on the nameplate?

A) 2 times the rated load current
B) 4 times the rated load current
C) 6 times the rated load current
D) 8 times the rated load current

Answer: C

For small motors with no nameplate rating, and which are not covered in NEC Tables 430.247, 430.248, 430.249, or 430.250, the locked-rotor current shall be assumed to be six times the rated load-current. NEC Section 440.12(C).

9-23 What is the maximum overcurrent protective device allowed on a 480-volt branch circuit with several small motors connected — all of which are under 1 horsepower and the maximum load of each motor is 5 amperes?

A) 10 amperes
B) 15 amperes
C) 20 amperes
D) 30 amperes

Answer: B

NEC Section 430.53(A) requires that a branch circuit of 600 volts, nominal, or less, containing several motors, each not exceeding 1 horsepower, must be provided with overcurrent protection not to exceed 15 amperes.

9-24 A 240-volt single-phase room air conditioner shall be considered as a single-phase motor unit if its rating is not more than:

A) 20 amperes
B) 30 amperes
C) 40 amperes
D) 50 amperes

Answer: C

NEC Section 440.62(A)(2) specifies that an air conditioner shall be considered a single motor unit if its rating is not more than 40 amperes and 250 volts, single phase.

9-25 Unless identified for a different value, motor control circuit devices with screw-type pressure terminals used with No. 14 AWG or smaller copper conductors must be torqued to a minimum pound-inch value. What is this value?

A) 2 pound-inches
B) 5 pound-inches
C) 7 pound-inches
D) 9 pound-inches

Answer: C

NEC Section 430.9(C) specifies a minimum of 7 pound-inches.

9-26 What must be used at all openings in an enclosure, conduit box, or barrier where wires pass through the opening in motor circuits?

A) A locknut

B) An approved metallic fitting

C) A bushing

D) A "red-head"

Answer: C

NEC Section 430.13 requires the use of a bushing.

9-27 A protective device for limiting surge voltages by discharging or bypassing surge current, and also preventing continued flow of follow current, is called a:

A) Surge arrestor

B) Auto fuse

C) Ground-fault circuit-interrupter

D) Surge bypass

Answer: A

NEC Section 280.2 covers surge arrestors and their use.

9-28 A phase converter is a device that:

A) Converts single-phase electrical power for the operation of equipment that normally operates from a three-phase electrical supply

B) Reverses the rotation of single-phase motors

C) Converts three-phase electrical power for the operation of equipment that normally operates from a single-phase electrical supply

D) Reverses the rotation of three-phase motors

Answer: A

The definition for a phase converter is given in NEC Section 455.2 and coincides with Answer A above.

9-29 When using a phase converter for a fixed load, the single-phase conductors must have an ampacity not less than what percentage of the full-load current rating of the motor or other load being served where the input and output voltages are the same?

A) 300%

B) 250%

C) 150%

D) 175%

Answer: B

NEC Section 455.6(A)(2) requires the conductors to be sized not less than 250% for fixed loads.

9-30 For continuous duty, the conductors connecting the secondary of a wound-rotor ac motor to its controller must have an ampacity not less that what percent of the full-load secondary current of the motor?

A) 100%

B) 110%

C) 125%

D) 130%

Answer: C

NEC Section 430.23(A) requires the conductors to have a current-carrying rating not less than 125% of the full-load secondary current of the motor.

9-31 If a motor uses a cord and plug as a disconnect, the attachment plug must be rated for:

A) Horsepower

B) Resistance

C) Voltage only

D) None of the above

Answer: A

NEC 430.109(F) requires the attachment plug to have a horsepower rating no less than the motor ratings.

9-32 If three three-phase motors are connected to a 120/240-volt single- to three-phase converter, and each motor has a full-load current rating of 8 amperes, what is the minimum amperage rating of the conductors feeding the phase converter?

A) 30.7 amperes

B) 60 amperes

C) 70.8 amperes

D) 81.5 amperes

Answer: B

Three motors at 8 amperes each give a total of 24 amperes. Therefore, 8 + 8 + 8 = 24 x 250% = 60 amperes. NEC Section 455.6(A)(2)

9-33 In a 3-wire, three-phase ac system feeding a motor, with one conductor grounded, where must fuses for motor overload protection be inserted in the circuit?

A) A fuse must be inserted in each ungrounded conductor and also in the grounded conductor

B) No fuses are required for motor overload protection

C) A fuse must be inserted only in each ungrounded conductor

D) None of the above answers are correct

Answer: A

NEC Section 430.36 requires fuses to be inserted in series with all conductors in this situation.

9-34 What other protection is required on motor overload relays and other devices that are not capable of opening short circuits or ground faults?

A) They must be protected by an autotransformer

B) They must be protected with plug fuses only

C) They must be protected with cartridge fuses only

D) They must be protected with overcurrent devices

Answer: D

NEC Section 430.40 requires such overload relays to be protected by fuses or circuit breakers with ratings or settings in accordance with NEC Section 430.52 or by a motor short-circuit protector. NEC 430.40 has been revised to include ground faults.

9-35 When two disconnecting means are utilized, one for disconnecting the motor and motor controller from the circuit and the other for disconnecting the motor control circuit from its power supply, where must these two disconnects be located?

A) Within sight of each other

B) Within 50 feet of each other

C) Immediately adjacent to each other

D) None of the above

Answer: C

NEC Section 430.74(A) requires that such disconnects be located immediately adjacent to each other.

9-36 By definition, what is a motor controller?

A) A device used to troubleshoot motors

B) A device or group of devices that serves to govern the electric power delivered to the apparatus to which it is connected

C) A device containing overcurrent protection for the motor

D) An assembly of enclosed sections having a common power bus, to purposely introduce a delay in tripping action

NEC CHANGE

Answer: B

The definition of "controller" is given in NEC Article 100, Definitions. Motor controllers often offer additional motor control such as reversing, jogging, etc.

9-37 Every motor controller must be capable of starting and stopping a motor. It must also be capable of:

A) Interrupting the stalled-rotor of the motor

B) Providing overcurrent protection

C) Detecting variances in the circuit voltage

D) Providing ground-fault protection

Answer: A

NEC Section 430.82(A) requires that each motor controller must be capable of starting and stopping the motor it controls as well as interrupting the stalled-rotor of the motor to help prevent damage to the motor windings.

9-38 An autotransformer motor starter must be so designed that it has at least:

A) An "ON" position; nothing else is required

B) Two starting positions

C) An "OFF" position

D) Three starting positions

Answer: C

NEC Section 430.82(B) stipulates that an autotransformer motor starter has an "OFF" position.

9-39 Other than the requirements stated in Answer C in Question 9-38, what other design considerations are required of an autotransformer?

A) It must be designed so that it cannot be de-energized without the proper key

B) It must be designed so that it cannot rest in the starting position

C) It must be designed so that it rests in the starting position at all times

D) All of the above

Answer: B

NEC Section 430.82(B) requires autotransformers to be designed so that they cannot rest in the starting position or in any position that will render the overload device in the circuit inoperative.

9-40 Motor-starter rheostats for dc motors operated from a constant voltage supply must be equipped with automatic devices that will interrupt the power supply before the speed of the motor falls to a certain level. What is this speed?

A) One-half its normal value

B) One-third its normal value

C) Three-fourths its normal value

D) Two-thirds its normal value

Answer: B

NEC Section 430.82(C)(2) requires automatic devices to interrupt the supply before the speed of the motor has fallen to less than one-third its normal value.

9-41 What is the minimum number of conductors a motor controller must open?

A) Only the grounded conductor

B) Enough ungrounded conductors to stop the motor

C) All ungrounded conductors

D) No conductors need to be opened

Answer: B

A motor controller need open only enough conductors to stop the motor. For example, in a single-phase, two-wire circuit, only one conductor needs to be opened to stop the motor. If the motor controller also serves as a disconnecting means, it must open all ungrounded conductors as stated in NEC Section 430.84, Exception.

9-42 How many motors connected to a circuit in excess of 600 volts are allowed to be connected to each motor controller?

A) 1

B) 2

C) 3

D) 4

Answer: A

NEC Section 430.87 requires each motor to be provided with an individual controller, with some exceptions.

9-43 Where one side of the motor control circuit is grounded, the motor control circuit should be arranged so that what type of ground in the remote control device will not start the motor?

A) An intentional ground

B) A low-voltage ground

C) An isolated ground

D) An accidental ground

Answer: D

NEC Section 430.73 requires that such control circuits be arranged so as to prevent accidental starting of the motor.

9-44 A disconnecting means for a motor or phase converter must plainly indicate:

A) The machines being fed

B) Whether it is in the open (OFF) or closed (ON) position

C) The phase rotation of the machine being fed

D) The date of installation

Answer: B

NEC Section 430.104 requires that all disconnects be plainly marked to show if the switch is on or off. In most cases, when the external handle is in the up position, the switch is closed and the circuit is energized. The switch must be marked "ON." When the operating handle is in the down position, the circuit is open and should be marked "OFF."

9-45 What must the ampere rating be for a disconnecting means for a motor circuit rated 600 volts, nominal, or less?

A) 100% of the motor nameplate current of the motor

B) At least 125% of the full-load current rating of the motor

C) At least 115% of the full-load current rating

D) At least 150% of the full-load current rating of the motor

Answer: C

NEC Section 430.110 requires the disconnecting means to be at least 115% of the full-load current rating of the motor. It can also be any amount above that rating; that is, it could be 125%, for example, and still comply with NEC regulations.

9-46 How many disconnecting means are required for electric motors receiving electrical energy from more than one source?

A) A disconnecting means is required for each source of electrical energy

B) One disconnect and a motor controller for each source of electrical energy

C) One disconnect must be provided for all sources

D) Two disconnects must be provided for each voltage source

Answer: A

NEC Section 430.113 requires a disconnect for each source of electrical energy. These disconnects must be mounted immediately adjacent to the equipment served. The code requires a permanent warning sign on or adjacent to each disconnecting means.

9-47 What are the jobs required of motor-circuit overcurrent protection?

A) Interrupt overloads

B) Interrupt fault currents

C) Both A and B

D) Neither A nor B

Answer: C

NEC Section 430.225(A) requires coordinated protection to automatically interrupt overload and fault currents in the motor, the motor circuit conductors, and the motor control apparatus.

9-48 Exposed live parts of motors and motor controllers operating at 50 volts or more between terminals must be guarded against accidental contact by a suitable enclosure, or as follows:

A) By installation on a suitable balcony or gallery that is inaccessible to unqualified personnel

B) By installation in a room or enclosure that is accessible only to qualified persons

C) By elevation of at least 8 feet above the floor

D) All of the above

Answer: D

NEC Section 430.232 recognizes all of the methods, A through C, as acceptable for protection from live motor parts.

9-49 When the frame of a stationary motor is not grounded and complies with *NEC* regulations, what provisions must be made?

A) It must be permanently and effectively insulated from the ground

B) It must be installed so that it can receive no energy from any source

C) The electrical energy must be less than 100 volts

D) The electrical energy must be less than 120 volts

Answer: A

NEC Section 430.242 permits ungrounded motor frames if the motor does not fall under items 1 through 4 of this NEC Section. Where the frame of the motor is not grounded, it must be permanently and effectively insulated from the ground.

9-50 The frames of all portable motors over a certain voltage must be guarded or grounded. What is this voltage?

A) 100 volts

B) 120 volts

C) 150 volts

D) 220 volts

Answer: C

NEC Section 430.243 requires that the frames of portable motors operating over 150 volts to ground be guarded or grounded.

9-51 What is the minimum size disconnecting means allowed for a hermetic refrigerant motor-compressor?

A) 100% of its nameplate rated-load current
B) 115% of its nameplate rated-load current
C) 125% of its nameplate rated-load current
D) 130% of its nameplate rated-load current

Answer: B

NEC Section 440.12(A)(1) requires the ampere rating of the disconnect to be at least 115% of the compressor's nameplate rated-load current or branch-circuit selection current, whichever is greater.

9-52 Where must the disconnecting means for an air-conditioning or refrigerating apparatus be located?

A) Within sight of the apparatus
B) Out of sight
C) At least 6 feet away from the equipment
D) At least 12 feet away from the equipment

Answer: A

NEC Section 440.14 requires that the disconnecting means for such equipment be located within sight of the equipment. NEC Article 100 also limits the maximum distance to 50 feet. The disconnecting means shall not be located on panels that are designed to allow access to the equipment.

9-53 Branch-circuit conductors supplying a single motor-compressor must have an ampacity not less than what percent of the motor-compressor rated-load current?

A) 100%
B) 115%
C) 125%
D) 150%

Answer: C

NEC Section 440.32 requires that branch-circuit conductors supplying a single motor-compressor must have an ampacity not less than 125% of either the motor-compressor rated-load current or the branch-circuit selection current, whichever is greater.

9-54 Each motor-compressor must be protected against overload and failure to start by which of the following means:

A) A separate overload relay that is responsive to motor-compressor current

B) A thermal protector integral with the motor-compressor

C) A fuse or inverse time circuit breaker

D) Any of the above

Answer: D

NEC Section 440.52 permits any of the methods listed in Answers A through C as an acceptable means of protecting motor-compressors against overloads and failures to start.

9-55 A room air conditioner is considered to be a single motor unit in determining its branch-circuit requirements when the following conditions are met:

A) It is cord- and plug-connected and its rating is not more than 40 amperes and 250 volts, single phase

B) The rating of the branch-circuit short-circuit and ground-fault protective device does not exceed the ampacity of the branch-circuit conductors or the rating of the receptacle, whichever is less

C) Total rated-load current is shown on the room air-conditioner nameplate rather than individual motor currents

D) All of the above

Answer: D

NEC Section 440.62(A) requires all of the conditions listed in Answers A through C to be met for the room air conditioner to be considered a single motor unit.

9-56 If a single motor-compressor has a nameplate rated-load current of 34.5 amperes and this is greater than the branch-circuit selection current, what should the current-carrying rating be for the branch-circuit conductors, rounded off to the closest whole number?

A) 36 amperes

B) 38 amperes

C) 43 amperes

D) 48 amperes

Answer: C

$34.5 \times 1.25 = 43.125$ amperes, rounded off to 43 amperes.

9-57 What is the minimum rating for a disconnecting means for the circuit in Question 9-56, rounded off to the closest whole number?

A) 30 amperes
B) 35 amperes
C) 40 amperes
D) 50 amperes

Answer: C

34.5 × 1.15 = 39.675 = 40 amperes, complying with NEC Section 440.12(A)(1).

9-58 If a replacement motor-compressor draws 38.7 amperes, what is the minimum rating for a disconnect means?

A) 44.5 amperes
B) 50.1 amperes
C) 60.7 amperes
D) 70 amperes

Answer: A

38.7 × 1.15 = 44.505. or 44.5 amperes. NEC Section 440.12(A)(1).

9-59 Which of the following are recognized types of phase converters?

A) Shaded-pole
B) Static and rotary
C) Compound-wound
D) Split-phase

Answer: B

NEC Section 455.2 describes two types of phase converters: static and rotary.

Chapter 10

Special Occupancies

Certain buildings or areas, because of their construction, use, or possible risk to persons or property, are termed special occupancies by the *NEC*. Compliance with special rules is required for these locations to assure the safe installation of electrical systems and their related equipment. Consequently, all electrician's exams will have some questions involving these special occupancies.

Wiring in Hazardous Locations

Articles 500 through 504 of the *NEC* cover the requirements of electrical equipment and wiring for all voltages in locations where fire or explosion hazards may exist. Locations are classified depending on the properties of the flammable vapors, liquids, gases, or combustible dusts or fibers that may be present, as well as the likelihood that a flammable or combustible concentration or quality is present.

Any area in which the atmosphere or a material in the area is such that the arcing of operating electrical contacts, components, and equipment may cause an explosion or fire is considered a hazardous location. In all such cases, explosion-proof equipment, raceways, and fittings are used to provide an explosionproof wiring system.

Hazardous locations have been classified in the *NEC* into certain class locations. Various atmospheric groups have been established on the basis of the explosive character of the atmosphere for the testing and approval of equipment for use in the various groups.

Garages and Similar Locations

Garages and similar locations where volatile or flammable liquids are handled or used as fuel in self-propelled vehicles (including automobiles, buses, trucks, and tractors) are not usually considered critically hazardous locations. However, the entire area up to a level 18 inches above the floor is considered a Class I, Division 2 location, and certain precautionary measures are required by the *NEC*. Likewise, any pit or depression below floor level shall be considered a Class I, Division 2 location, and the pit or depression may be judged as Class I, Division 1 location if it is unvented.

Normal raceway (conduit) and wiring may be used for the wiring method above this hazardous level, except where conditions indicate that the area concerned is more hazardous than usual. In this case, the applicable type of explosionproof wiring may be required.

Approved seal-off fittings should be used on all conduit passing from hazardous areas to nonhazardous areas. The requirements set forth in *NEC* Sections 501.15 and 501.15(B)(2) shall apply to horizontal as well as vertical boundaries of the defined hazardous areas. Raceways embedded in a masonry floor or buried beneath a floor are considered to be within the hazardous area above the floor if any connections or extensions lead into or through such an area. However, conduit systems terminating to an open raceway, in an outdoor unclassified area, shall not be required to be sealed between the point at which the conduit leaves the classified location and enters the open raceway.

Airport Hangars

Buildings used for storing or servicing aircraft in which gasoline, jet fuels, or other volatile flammable liquids or gases are used fall under Article 513 of the *NEC*. In general, any pit or depression below the level of the hangar floor is considered to be a Class I, Division 1 location. The entire area of the hangar including any adjacent and communicating area not suitably cut off from the hangar is considered to be a Class I, Division 2 location up to a level 18 inches above the floor. The area within 5 feet horizontally from aircraft power plants, fuel tanks, or structures containing fuel is considered to be a Class I, Division 2 hazardous location; this area extends upward from the floor to a level 5 feet above the upper surface of wings and engine enclosures.

Adjacent areas in which hazardous vapors are not likely to be released, such as stock rooms and electrical control rooms, should not be classed as hazardous when they are adequately ventilated and effectively cut off from the hangar itself by walls or partitions. All fixed wiring in a hangar not within a hazardous area as defined in Section 513.3 must be installed in metallic raceways or shall be Type MI or Type ALS cable. The only exception is wiring in nonhazardous locations as defined in Section 513.3(D), which may be of any type recognized in Chapter 3 (Wiring Methods and Materials) in the *NEC*.

Theaters

The *NEC* recognizes that hazards to life and property due to fire and panic exist in theaters, cinemas, and the like. The *NEC* therefore requires certain precautions in these areas in addition to those for commercial installations.

These requirements include the following:

- Proper wiring of motion picture projection rooms (Article 540)
- Heat-resistant, insulated conductors for certain lighting equipment (Section 520.42)
- Adequate guarding and protection of the stage switchboard and proper control and overcurrent protection of circuits (Section 520.22)
- Proper type and wiring of lighting dimmers (Sections 520.53(E) and 520.25)
- Use of proper types of receptacles and flexible cables for stage lighting equipment (Section 520.45)
- Proper stage flue damper control (Section 520.49)
- Proper dressing-room wiring and control (Sections 520.71, .72, and .73)
- Fireproof projection rooms with automatic projector port closures, ventilating equipment, emergency lighting, guarded work lights, and proper location of related equipment (Article 540).

Outdoor or drive-in motion picture theaters do not present the inherent hazards of enclosed auditoriums. However, the projection rooms must be properly ventilated and wired for the protection of the operating personnel. You can rest assured that questions about these areas will be asked on all electrician's exams.

Hospitals

Hospitals and other health-care facilities fall under Article 517 of the *NEC*. Part II of Article 517 covers the general wiring of health-care facilities, including the performance criteria and wiring methods to minimize shock hazards to patients in electrically susceptible patient areas. Part III covers essential electrical systems for hospitals. Part IV covers the requirements for electrical wiring and equipment used in inhalation anesthetizing locations.

With the widespread use of X-ray equipment of varying types in health-care facilities, electricians are often required to wire and connect equipment such as discussed in Article 660 of the *NEC*. Conventional wiring methods are used, but provisions should be made for 50- and 60-ampere receptacles for medical X-ray equipment.

Anesthetizing locations of hospitals are deemed to be Class I, Division 1, to a height of 5 feet above floor level. Gas storage rooms are designated as Class I, Division 1, throughout.

The *NEC* recommends that wherever possible electrical equipment for hazardous locations should be located in less hazardous areas. It also suggests that by adequate, positive-pressure ventilation from a clean source of outside air, the hazards may be reduced or hazardous locations limited or eliminated. In many cases the installation of dust-collecting systems can greatly reduce the hazards in a Class II area.

10-1 Conduit in hazardous locations must be made up:

A) Sealed
B) Bonded
C) Wrenchtight
D) Snug

Answer: C

All conduit used in hazardous locations must be threaded with a NPT standard conduit cutting die that provides 3/4-inch taper per foot. Such conduit must be made up wrenchtight to minimize sparking when fault current flows through the conduit system. NEC Section 500.8(D).

10-2 Locations where easily ignitable fibers are stored are:

A) Class I, Division 1
B) Class II, Division 1
C) Class III, Division 1
D) Class III, Division 2

Answer: D

A Class III, Division 2 location is a location in which easily ignitible fibers are stored or handled other than in the process of manufacture. NEC Section 500.5(D)(2).

10-3 All fixed boxes, fittings and joints shall be made explosionproof in:

A) Class I, Division 1 locations
B) Class II, Division 2 locations
C) Class II, Division 1 locations
D) Class III, Division 1 locations

Answer: A

In Class I, Division 1 locations, all such conduit connections shall be explosionproof. In addition, all threaded connections shall have at least five threads fully engaged. NEC Section 500.7(A).

10-4 The disconnecting means for a 120/240-volt circuit for a gasoline dispenser must disconnect:

A) The neutral
B) All conductors
C) The grounded conductor
D) The ungrounded conductors

Answer: B

Each circuit leading to or through a dispensing pump must be provided with a switch or other acceptable means to disconnect simultaneously from the source of supply all conductors of the circuit, including the grounded conductor, if any. NEC Section 514.11(A).

10-5 In Class I, Division 1 locations, a conduit to a splice box must be sealed if the trade size is:

A) 1 inch or larger
B) 2½ inches or larger
C) 2 inches or larger
D) ½ inch or larger

Answer: C

In each conduit of 2-inch size or larger entering an enclosure or fitting housing terminals, splices or taps must contain a seal within 18 inches of such enclosure or fitting. NEC Section 501.15(A)(1).

10-6 The external surface of totally enclosed motors of Types (2) or (3) shall never have an operating temperature (in degrees Celsius) in excess of what percentage of the ignition temperature of the gas or vapor involved in its operation?

A) 75%
B) 80%
C) 100%
D) None of these

Answer: B

Furthermore, approved devices shall detect and automatically de-energize the motor if the temperature of the motor exceeds the designed limits. NEC Section 501.125(B).

10-7 In a Class I location, a pendant fixture with a rigid stem longer than 12 inches must be braced:

A) 12 inches from the top

B) Not more than 12 inches from the fixture

C) 16 inches from the top

D) On the stem

Answer: B

For stems longer than 12 inches, permanent and effective bracing against lateral displacement shall be provided at a level not more than 12 inches above the lower end of the stem. NEC Section 501.130(A)(3).

10-8 A dust-ignitionproof enclosure supplied by a horizontal raceway does not require a raceway seal if the conduit length is at least:

A) 10 feet

B) 5 feet

C) 8 feet

D) 12 feet

Answer: A

Where a raceway provides communication between an enclosure that is required to be dust-ignitionproof and one that is not, suitable means shall be provided to prevent the entrance of dust into the dust-ignitionproof enclosure through the raceway. One of the following means shall be permitted: (1) a permanent and effective seal; (2) a horizontal raceway not less than 10 feet long; or (3) a vertical raceway not less than 5 feet long and extending downward from the dust-ignitionproof enclosure. NEC Section 502.15.

10-9 Motors in Class II, Division 1 locations must be identified for Class II, Division 1 locations, or be:

A) Totally enclosed

B) Totally enclosed pipe-ventilated

C) Explosionproof

D) Air ventilated

Answer: B

In Class II, Division 1 locations, motors, generators, and other rotating electrical machinery shall be totally enclosed pipe-ventilated, meeting temperature limitations in NEC 502.5. NEC Section 502.125(A).

10-10 When a flexible cord supplies a pendant fixture in a Class II location, it must be listed for:

A) Class II
B) Extra-hard usage
C) Hard usage
D) Normal usage

Answer: B

Where necessary to employ flexible connections . . . a flexible cord listed for extra-hard usage and suitable seals shall be used. NEC Section 502.140(A)(1).

10-11 In Class III locations, pendant fixtures must be braced or have a flexible connector if the stem is longer than:

A) 6 inches
B) 10 inches
C) 24 inches
D) 12 inches

Answer: D

For stems longer than 12 inches, permanent and effective bracing against lateral displacement must be provided at a level not more than 12 inches above the lower end of the stem, or provided not more than 12 inches from the point of attachment to the supporting box or fitting. NEC Section 501.130(A)(3).

10-12 In a garage, unenclosed equipment that may produce arcs or sparks must be placed at least what height above the floor?

A) 10 feet
B) 18 inches
C) 16 feet
D) 12 feet

Answer: D

Equipment that is less than 12 feet above the floor level and that may produce arcs, sparks, or particles of hot metal, such as cutouts, switches, charging panels, generators, motors, or other equipment . . . shall be of the totally enclosed type or so constructed as to prevent escape of sparks or hot metal particles. NEC Section 511.7(B).

10-13 Which circuit is one in which a spark or thermal effect is incapable of causing ignition of a mixture of flammable or combustible material in air under prescribed test conditions?

A) Intrinsically safe
B) Low voltage
C) Hazard-proof
D) Explosion-proof

Answer: A

The NEC defines an intrinsically safe circuit in Section 504.2.

10-14 The hazardous area near a gasoline fill-pipe from an underground tank with loose fill extends horizontally:

A) 10 feet
B) 20 feet
C) 5 feet
D) 6 feet

Answer: A

At the fill opening for an underground tank, Class I, Division 2 extends up to 18 inches above grade level within a horizontal radius of 10 feet from a loose fill connection and within a horizontal radius of 5 feet from a tight fill connection. NEC Table 514.3(B)(1).

10-15 The Class I, Division 2 location in a remote outdoor gasoline dispensing device extends from the base upward for:

A) 18 inches
B) 3 feet
C) 4 feet
D) 6 feet

Answer: A

NEC Table 514.3(B)(1) states that Class I, Division 2 is within 3 feet of any edge of a pump, extending in all directions, and up to 18 inches above floor or grade level within 10 feet horizontally from any edge of the pump.

10-16 Equipment used in hazardous (classified) locations must be marked to show the Class, Group, and operating temperature or temperature class referenced to a:

A) 30° C ambient

B) 40° C ambient

C) 50° C ambient

D) 60° C ambient

Answer: B

NEC Section 500.8(B) requires the markings to be based on 40° C ambient.

10-17 A location in which flammable gases or vapors are or may be present in the air in quantities sufficient to produce explosive or ignitible mixtures is classified as a:

A) Class IV location

B) Class III location

C) Class II location

D) Class I location

Answer: D

NEC Section 500.5(B) gives this definition as a Class I location.

10-18 A hazardous location in which ignitible concentrations of flammable gases or vapors can exist under normal operating conditions is classified as a:

A) Class I, Division 1 location

B) Class I, Division 2 location

C) Class II, Division 1 location

D) Class II, Division 2 location

Answer: A

NEC Section 500.5(B)(1) specifies the above condition to be Class I, Division 1.

10-19 A hazardous location in which ignitible concentrations of gases or vapors may exist frequently because of repair or maintenance operations or because of leakage is classified as a:

A) Class I, Division 1 location

B) Class I, Division 2 location

C) Class II, Division 1 location

D) Class II, Division 2 location

Answer: A

NEC Section 500.5(B)(1) also specifies the conditions in this question as Class I, Division 1.

10-20 A hazardous location in which breakdown or faulty operation of equipment or processes might release ignitible concentrations of flammable gases or vapors, and might also cause simultaneous failure of electric equipment is classified as a:

A) Class I, Division 1 location

B) Class I, Division 2 location

C) Class II, Division 1 location

D) Class II, Division 2 location

Answer: A

NEC Section 500.5(B)(1) also specifies the conditions in this question as Class I, Division 1.

10-21 When using flexible metal conduit or liquidtight flexible metal conduit in hazardous areas where permitted, what must be provided to ensure equipment grounding?

A) The metal jacket on the flexible metal conduit is sufficient

B) A ground rod driven at each piece of equipment

C) A separate bonding jumper

D) All of the above

Answer: C

NEC Section 501.30(B) requires that such installations be provided with internal or external bonding jumpers in parallel with each conduit and complying with NEC Section 250.102.

10-22 When surge arresters are installed in hazardous locations, what type of enclosure must be provided?

A) A PVC (plastic) junction box

B) An enclosure identified for the location

C) 4 x 4 junction box

D) A handy box

Answer: B

NEC Section 502.35 calls for surge arresters to be installed in an enclosure identified for the location. Furthermore, surge-protective capacitors must be of a type designed for specific duty.

10-23 When installing 480-volt dry-type transformers in Class II, Division 2 locations, the transformers must be installed in vaults unless:

A) Their windings and terminal connections are enclosed in a tight metal housing and operating at 600V or less.

B) They are rated above 10 kVA

C) They are rated below 10 kVA

D) They are enclosed in a properly ventilated housing

Answer: A

NEC Section 502.100(B)(3) does not require a vault if the conditions in answer A are met.

10-24 Where flexible cords are used in Class II, Division 1 locations, they must:

A) Have THHN insulation

B) Be listed for extra-hard usage

C) Be approved for hard usage

D) Have TW insulation

Answer: B

NEC Section 502.140 requires that such cords be listed for extra-hard usage.

10-25 Flexible cord connectors used in Class II, Division 2 locations are required to have:

A) A drip pan

B) A warning label

C) Suitable seals

D) An orange tape at each termination point

Answer: C

NEC Section 502.140 requires that suitable seals be provided to eliminate the entrance of dust particles.

10-26 When resistors and resistance devices are installed in Class II locations, what type of enclosures must be provided for them?

A) Dustproof only

B) Ventilated

C) Waterproof

D) Dust-ignitionproof

Answer: D

NEC Section 502.120(B)(3) requires the enclosures to be dust-ignitionproof identified for Class II locations.

10-27 Besides rigid metal conduit, rigid nonmetallic conduit, intermediate metal conduit, and EMT, which of the following cables are allowed in Class III, Division 1 locations?

A) Type NM cable

B) Type AC cable

C) Type MI cable

D) Type SE cable

Answer: C

Besides the conduits listed in the question, NEC Section 503.10(A) allows dusttight wireways, Type MI cable and Type MC cable with listed termination fittings.

10-28 Lighting fixtures installed in Class III, Divisions 1 and 2 that may be exposed to physical damage must be protected by:

A) Using lower wattage lamps

B) A suitable guard

C) Smaller lamps

D) By elevating the lamps 8 feet or more above floor level

Answer: B

NEC Section 503.130(B) requires that such lighting fixtures be protected with a suitable guard. Suitable guards usually consist of cast-iron shields over a hardened glass enclosure.

10-29 When installing conduit in Class III locations, the locknut-bushing and double-locknut types of contacts shall:

A) Be used for all such installations

B) Be approved for bonding purposes if fitted with a fiber bushing

C) Not be depended upon for bonding purposes

D) Used exclusively on rigid metal conduit for bonding purposes

Answer: C

NEC Section 503.30(A) states that locknut-bushing and double-locknut contacts shall not be depended upon for bonding purposes. Rather, a bonding jumper must be used.

10-30 Each floor level in a commercial garage, up to a level of 18 inches above the floor, is considered to be:

A) Class I, Division 1

B) Class I, Division 2

C) Class II, Division 1

D) Class II, Division 2

Answer: B

NEC Section 511.3(B)(2) defines the area in the question as a Class I, Division 2 location.

10-31 Any pit or depression below floor level in a commercial garage is classified as:

A) Class I, Division 1

B) Class I, Division 2

C) Class II, Division 1

D) Class II, Division 2

Answer: A

NEC Section 511.3(B)(3) classifies such areas as Class I, Division 1 areas except where the area has six air changes per hour; it may then be classified as Class I, Division 2.

10-32 Areas adjacent to main commercial garage areas, such as mechanical rooms or stock rooms, are not classified as hazardous areas if proper ventilation is provided. What is the minimum air changes per hour for these areas not to be so classified?

A) One air change per hour

B) Two air changes per hour

C) Three air changes per hour

D) Four air changes per hour

Answer: D

NEC Section 511.3(A)(3) requires four air changes per hour or the area must be effectively cut off by walls or partitions.

10-33 Which of the following wiring methods are allowed for fixed wiring above Class I locations?

A) Type MI cable

B) Type TC cable

C) Type MC cable

D) All of the above

Answer: D

NEC Section 511.7(A)(1) allows metallic raceways, rigid nonmetallic conduit, electrical nonmetallic tubing, Type MI, Type TC, Type MC or PLTC cable. This article has been revised to permit ITC cable to be used in accordance with Article 727.

10-34 Motors installed above Class I locations, but less than 12 feet above floor level and that may produce sparks or hot metal particles, must be of what type?

A) Ventilated

B) Open squirrel-cage motors with cast-iron guards

C) Totally enclosed

D) None of the above

Answer: C

NEC Section 511.7(B)(1) calls for totally enclosed motors or else so constructed as to prevent escape of sparks or hot metal particles. This section has been re-titled "Electrical Equipment Installed Above Class I Locations" to make the NEC more user-friendly.

10-35 What must be provided on all 125-volt (110-120 volt) circuits that are installed in commercial garages where electrical automotive diagnostic equipment, electrical hand tools, or portable lighting devices are to be used?

A) Seal-offs

B) Ground-fault circuit-interrupter protection for personnel

C) A continuous bonding jumper

D) None of the above

Answer: B

NEC Section 511.12 requires ground-fault circuit-interrupter protection for personnel.

10-36 Any pit or depression below floor level of an aircraft hangar is classified as:

A) Class I, Division 1 or Zone 1

B) Class I, Division 2 or Zone 2

C) Class II, Division 1

D) Class II, Division 2

Answer: A

NEC Section 513.3(A) classifies any pit or depression in an aircraft hangar, up to the floor level, as Class I, Division 1. The NEC now allows the same location to be identified as Zone 1.

10-37 The entire area of an aircraft hangar, including any adjacent and communicating areas not suitably cut off from the hangar, shall be classified as a Class I, Division 2 or Zone 2 location up to a level of how many inches above the floor?

A) 6 inches

B) 8 inches

C) 12 inches

D) 18 inches

Answer: D

NEC Section 513.3(B) classifies an area from the floor up to 18 inches (457 mm) above the floor level as a Class I, Division 2 or Zone 2 location.

10-38 Areas within 5 feet horizontally from aircraft power plants or aircraft fuel tanks are classified as Class I, Division 2 or Zone 2 locations. How far vertically does this area extend?

A) 5 feet above finished floor

B) 5 feet above cockpit

C) 5 feet above the upper surface of wings and of engine enclosures

D) 5 feet above propeller blades

Answer: C

NEC Section 513.3(C) classifies areas in the vicinity of aircraft 5 feet horizontally and 5 feet vertically above the upper surface of aircraft wings and of engine enclosures, whichever is greater.

10-39 Which of the following fixed wiring systems is *not* allowed in aircraft hangars outside Class I locations?

A) Metallic raceways

B) Type MI cable

C) Type AC cable

D) Type TC cable

Answer: C

NEC Section 513.7(A) allows metallic raceways, Type MI, TC, or Type MC cable for fixed wiring in areas outside of Class I locations in aircraft hangars. The Section does not allow Type AC cable.

10-40 When an aircraft is stored in a hangar, what provisions must be provided for the aircraft's electrical system?

A) It must remain energized

B) Circuit breakers must be closed

C) It must be arranged to charge all batteries

D) It must be de-energized

Answer: D

NEC Section 513.10(A)(1) requires any aircraft that is stored in a hangar to have its electrical system de-energized. Furthermore, the electrical system should also be de-energized, whenever possible, when the aircraft is undergoing maintenance.

10-41 How far above floor level must all electric equipment on aircraft energizers be located?

A) 8 inches

B) 10 inches

C) 12 inches

D) 18 inches

Answer: D

NEC Section 513.10(C)(1) requires that external power sources for energizing aircraft must be designed and mounted so that all electric equipment and fixed wiring will be at least 18 inches above floor level.

10-42 Flexible cords for aircraft energizers must be approved for the type of service and be rated for:

A) Hard usage

B) Extra-hard usage

C) Medium duty

D) Light duty

Answer: B

NEC Section 513.10(C)(3) requires that all cords and connectors for energizers used in aircraft hangars be identified for extra-hard usage.

10-43 The cords mentioned in Question 10-42 must also contain:

A) An equipment grounding conductor

B) A separate bonding jumper

C) At least two equipment grounding conductors

D) All of the above

Answer: A

NEC Section 513.10(C)(3) states ". . . and shall include an equipment grounding conductor." This will ensure that all equipment connected to the cord will be properly grounded, provided the correct plug or connector is used.

10-44 Which of the following may not be used for fixed wiring in bulk storage plants above Class I locations?

A) Metal raceways

B) PVC Schedule 80 rigid nonmetallic conduit with proper joints

C) Type SE cable

D) Type MI cable

Answer: C

NEC Section 515.7(A) allows metallic raceways, PVC Schedule 80 rigid nonmetallic conduit, Type MI, Type TC, or Type MC cable. It does not allow Type SE cable to be used.

10-45 The interior of paint spray booths and rooms are classified as which hazardous area, according to the *NEC*?

A) Class I or Class II, Division 1

B) Class I or Class II, Division 2

C) Class I or Class II, Division 3

D) Class I or Class II, Division 4

Answer: A

NEC Section 516.3(B) lists the interior of spray booths and rooms except as specifically provided in NEC Section 516.3(D) as Class I or Class II, Division 1 locations.

10-46 Adjacent areas to paint spray booths that are cut off from Class I or Class II locations by tight partitions without communicating openings are classified as:

A) Class I

B) Hazardous areas

C) Division 3 areas

D) Unclassified areas

Answer: D

Areas as described in this question, which contain no hazardous vapors or fumes, are classified as unclassified areas according to NEC Section 516.3(E).

10-47 Which one of the following wiring methods is approved for fixed wiring above Class I or Class II locations in spray booth areas?

A) Type AC cable

B) Type NM cable

C) Type SE cable

D) Type MI cable

Answer: D

NEC Section 516.7(A) requires that all wiring above Class I or Class II locations in spray booth areas be in either metallic conduit, EMT, rigid nonmetallic conduit, or shall be Type MI, TC, or MC cable. Type AC, NM, and SE cables are not allowed.

10-48 What means of grounding the terminals of receptacles (100 volts and over) is required in areas used for patient care in health care facilities?

A) An insulated aluminum wire

B) A bare copper conductor

C) An insulated copper conductor

D) A bare aluminum conductor

Answer: C

Grounding terminals of receptacles and all noncurrent-carrying conductive surfaces of fixed electric equipment likely to become energized and subject to personal contact, must be grounded by an insulated copper conductor. This conductor must be sized in accordance with NEC Table 250.122 and installed in metal raceways or metal-clad cables with the branch-circuit conductors supplying the receptacles. NEC Section 517.13(B).

10-49 The panelboard serving patient care areas must have the equipment grounding terminal busses bonded together with an insulated continuous copper conductor not smaller than:

A) No. 12 AWG

B) No. 10 AWG

C) No. 8 AWG

D) No. 6 AWG

Answer: B

NEC Section 517.14 calls for an insulated copper conductor not smaller than No. 10 AWG for this purpose.

10-50 What is the minimum number of branch circuits allowed at each patient bed location?

A) 1
B) 2
C) 3
D) 4

Answer: B

NEC Section 517.18(A) requires at least two branch circuits at each patient bed location, one from the emergency system and one from the normal system.

10-51 What is the minimum number of receptacles required at each patient bed location in critical care areas of health care facilities?

A) 2
B) 4
C) 6
D) 8

Answer: C

NEC Section 517.19(B) requires at least six hospital grade receptacles fed by a minimum of two branch circuits. Three duplex receptacles would meet this requirement.

10-52 Circuits classified as life safety branch and critical branch of the emergency electrical system in health care facilities must be:

A) Installed in the same raceways with normal branch circuits
B) Installed parallel with normal circuits
C) Kept entirely independent of all other wiring and equipment
D) None of the above

Answer: C

NEC Section 517.30(C) requires that the emergency wiring system be kept entirely independent of all other wiring and equipment and must not enter the same raceway, boxes or cabinets with each other or other wiring except in transfer switches.

10-53 What is the minimum number of power sources for essential electrical systems in health care facilities?

A) 1

B) 2

C) 3

D) 4

Answer: B

NEC Section 517.35 requires at least two independent sources of power for these circuits in case one power source should fail.

10-54 Which of the following is not approved as an alternate power source for essential electrical systems in health care facilities?

A) Generator

B) Self-contained battery integral with the equipment

C) Battery system

D) Solar powered equipment

Answer: D

NEC Section 517.45 permits the use of A, B, and C above, but not D. Obviously, D would offer no energy at night or on cloudy days.

10-55 Where must the alternate power for a health care facility be located?

A) 2 feet from main

B) On premises

C) In vaulted ceiling

D) Remote location

Answer: B

NEC Section 517.44(B) requires alternate power to be located on premises. Section 517.44(C) requires that careful consideration be given to the location that houses these components to minimize interruptions due to storms, floods, earthquakes, etc.

10-56 In a health care facility, receptacles supplied from the emergency system shall?

A) Not be utilized

B) Have distinctive markings

C) Have 30 amp ratings

D) Be installed in hallways and bathrooms

Answer: B

NEC Section 517.41(E) requires that either the receptacles or plates be color coded to make them readily identifiable.

Chapter 11

Miscellaneous Applications

The *NEC* specifies rules for certain special electrical equipment which supplement or modify the general *NEC* rules. Some of the items falling under this category include: electric signs and outline lighting, manufactured wiring systems, office furnishings, cranes and hoists, elevators, dumbwaiters, escalators, moving walks, electric welders, induction and dielectric heating equipment, industrial machines, and similar applications.

The wiring methods, types of lampholders, and type of transformers used in the construction of a sign are defined in detail by the *NEC* according to the range of operating voltage of the lamps used. In general, signs operating at 600 volts or less use either incandescent lamps or electric-discharge lamps. In most cases, signs operating at over 600 volts employ electric-discharge lamps only and the *NEC* rules are more extensive because of the increased fire and shock hazard associated with high voltages (up to 15,000 volts).

Control wiring and feeder connection terminals on nonportable, electrically driven machines are usually installed at the factory. In most cases, due to the areas in which the equipment is used, the wiring method is restricted to rigid conduit except for short lengths of flexible conduit where necessary for final connection to the equipment. Continuously moving parts of the machine are interconnected with approved type, extra flexible, nonmetallic covered cable. The size of the conductors, type of mounting of control equipment, overcurrent protection, and grounding are covered in Article 670 of the *NEC*.

The electric supply for metalworking machines may be from conventional branch circuits or feeders or in the form of bus ducts or wireways. These two latter methods provide a very flexible type of installation allowing the moving of machines from one part of the plant or shop to another. Their reconnection to another part of the bus duct system is almost instantaneous, eliminating changes in the raceway wiring. Bus duct systems are covered in Article 368 of the *NEC*.

The wiring for, and connection of, induction and dielectric heat generating equipment used in industrial and scientific applications (but not for medical or dental applications) are covered in Article 665 of the *NEC*.

The heating effect of such equipment is accomplished by placing the materials to be heated in the magnetic field of an electric voltage of very high frequency or between two electrodes connected to a source of high frequency voltage. Induction heating is used in heating metals and other conductive materials. Dielectric heating is used in the heating of materials that are poor conductors of electric current.

The equipment used consists either of motor-operated, high-frequency generators, or electric tube or solid-state oscillators. Such equipment is supplied by manufacturers or their representatives. Designers, electrical contractors, and electrical workers can benefit by contacting these manufacturers to obtain installation procedures, specifications, and the like.

The use of computers more than doubles each year. Consequently data-processing rooms or areas are becoming more common. Article 645 of the *NEC* covers the installation of power supply wiring, grounding of equipment, and other such provisions that will insure a safe installation.

In many computer installations, the equipment operates continuously. For this reason, Section 645.5(A) of the *NEC* requires that branch circuits supplying one or more computers have an ampacity not less than 125% of the total connected load.

Separate data processing units shall be permitted to be interconnected by means of cables and cable assemblies listed for the purpose. Where run on the surface of the floor, cables shall be protected against physical damage.

Since there are so many interconnections for power, control, and communications, computer equipment is sometimes installed on raised floors with the cables running underneath the floor. When the wiring is installed beneath raised floors in computer rooms, branch-circuit conductors must be installed in either rigid conduit, intermediate metal conduit, electrical metallic tubing, metal wireway, surface metal raceway with metal cover, flexible metal conduit, liquidtight flexible metal or nonmetallic conduit, Type MI cable, Type MC cable, or Type AC cable. The *NEC* has added electrical nonmetallic tubing and nonmetallic surface raceways as approved wiring methods. Whichever wiring method is used, it must be installed in accordance with Section 300.11.

A disconnecting means must be provided that will allow the operator to disconnect all computer equipment in the area. A disconnecting means must also be provided to disconnect the ventilation system serving the computer area.

Electric welding equipment is normally treated as a piece of industrial power equipment for which branch circuits adequate for the current and voltage of the equipment are provided. Certain specific conditions, however, apply to circuits feeding ac/dc transformers and DC rectifier arc welders, motor-generator arc welders, resistance welders, and the like. These requirements are found in Article 630 of the *NEC*.

Crane and hoist equipment is usually furnished and mechanically installed by crane manufacturing companies or their representatives. When working on such equipment, refer to Article 610.

The *NEC* recognizes the potential danger of electric shock to persons in swimming pools, wading pools, and therapeutic pools, or near decorative pools or fountains. Accordingly, the *NEC* provides rules for the safe installation of electrical equipment and wiring in or adjacent to swimming pools and similar locations. Article 680 of the *NEC* covers the specific rules governing their installation and maintenance.

Other installations falling under the category of special equipment are listed below along with the appropriate *NEC* Article for further reference.

- Electrically driven or controlled irrigation machines, Article 675
- Electrolytic cells, Article 668
- Electroplating, Article 669
- Elevators, dumbwaiters, escalators, moving walks, wheelchair lifts and stairway chair lifts, Article 620
- Integrated electric systems, Article 685
- Manufactured wiring systems, Article 604
- Office furnishings, Article 605
- Pipe organs, Article 650
- Solar photovoltaic systems, Article 690
- Sound-recording and similar equipment, Article 640
- X-ray equipment, Article 660

11-1 Switches or similar devices controlling a transformer in a sign must have a rating based on what percentage of the transformer's ampere rating?

A) 100%

B) 200%

C) 125%

D) 300%

Answer: B

NEC Section 600.6(B) requires that such devices have an ampere rating of not less than twice the ampere rating of the transformer.

11-2 A branch circuit supplying fluorescent lamps in a sign must not be rated more than:

A) 20 amperes

B) 15 amperes

C) 30 amperes

D) 40 amperes

Answer: A

Circuits supplying incandescent, fluorescent, and high-intensity-discharge lamp forms must be rated not more than 20 amperes. NEC Section 600.5(B)(1).

11-3 Wood used for decoration on a sign must not be closer to a lampholder than:

A) 6 inches

B) 1 foot

C) 3 inches

D) 2 inches

Answer: D

Wood is permitted on signs for decoration if placed not less than 2 inches from the nearest lampholder or current-carrying part. NEC Section 600.9(C).

11-4 What is the maximum length allowed for cords supplying portable signs in a dry location?

A) 16 feet

B) 12 feet

C) 15 feet

D) 8 feet

Answer: C

No cord used for portable signs in a dry location shall be more than 15 feet in length. NEC Section 600.10(D).

11-5 What is the smallest size conductor permitted for wiring neon secondary circuits rated at 1000 volts or less?

A) No. 14 AWG

B) No. 12 AWG

C) No. 16 AWG

D) No. 18 AWG

Answer: D

NEC Section 600.31(B) restricts the wire size to No. 18 AWG or larger when the circuit voltage does not exceed 1000 volts.

11-6 In wet locations, which of the following must be used to close the opening between neon tubing and a receptacle?

A) A listed cap

B) Silicon sealant

C) Electrical tape

D) None of the above

Answer: A

NEC Section 600.42(F) requires a listed cap for closing the opening between neon tubing and a receptacle.

11-7 The installation of equipment having an open current voltage exceeding 1000 volts is *not* allowed in which of the following occupancies?

A) A commercial supermarket or grocery store

B) A multifamily dwelling

C) An industrial iron works

D) A shopping mall parking lot

Answer: B

NEC Section 600.32(I) does not permit open current voltage exceeding 1000 volts in any dwelling occupancies.

11-8 Which of the following wiring methods is suitable for underfloor wiring in a computer room?

A) Rigid conduit

B) Electrical metallic tubing

C) Electrical nonmetallic tubing

D) All of the above

Answer: D

NEC Section 645.5(D)(2) permits the use of rigid conduit, electrical metallic and nonmetallic tubing as well as other methods for this application.

11-9 Lighting fixtures above a pool without GFCI protection must be installed at a minimum height of:

A) 5 feet

B) 2 feet

C) 10 feet

D) 12 feet

Answer: D

Lighting fixtures without GFCI protection must not be installed over a pool or over the area extending 5 feet horizontally from the inside walls of a pool unless they are 12 feet above the maximum water level. NEC Section 680.43(B)(1). See Figure 11-1 on the next page.

11-10 In general, overhead conductors must not pass directly over a pool or horizontally within how many feet of the inside walls of the pool?

A) 5 feet

B) 18 feet

C) 10 feet

D) 3 feet

Answer: C

Overhead conductors must not be installed above a pool or the area extending 10 feet horizontally from the inside of the walls of the pool. NEC Section 680.8. See Figure 11-1 on the next page.

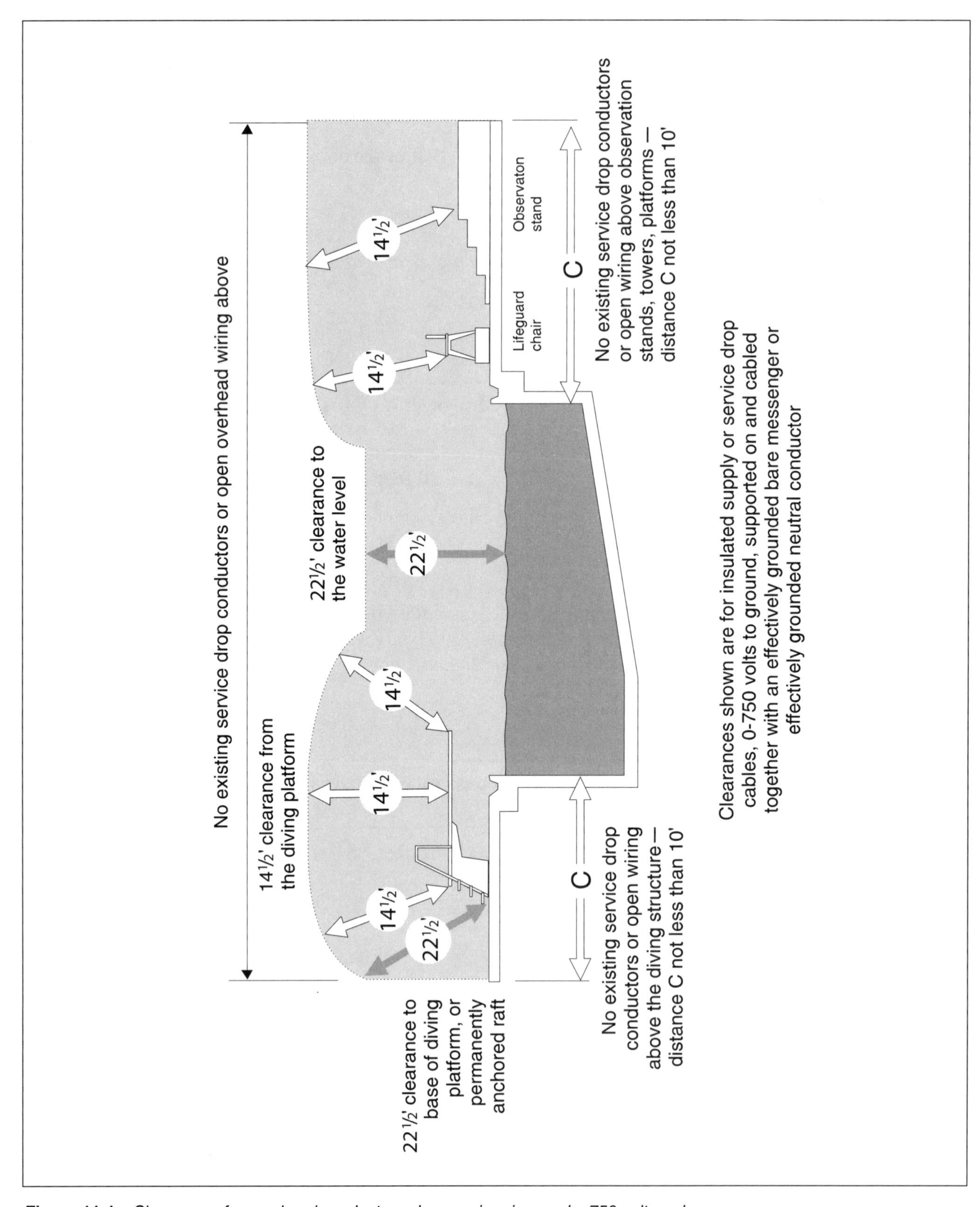

Figure 11-1: *Clearances for overhead conductors above swimming pools, 750 volts or less*

11-11 Utility-owned low-voltage communication, radio and television coaxial cables may pass directly over a pool if at a minimum height of:

A) 18 feet
C) 16 feet
B) 14 feet
D) 10 feet

Answer: D

Communication, radio and television coaxial cables within the scope of Articles 800-820, shall be permitted at a height of not less than 10 feet above swimming and wading pools, diving structures and observation stands, towers or platforms. NEC Section 680.8(B).

11-12 A typical wet-niche fixture must be installed below the water line at least:

A) 12 inches
C) 18 inches
B) 15 inches
D) 24 inches

Answer: C

Lighting fixtures mounted in walls of swimming pools must be installed with the top of the fixture lens at least 18 inches below the normal water level of the pool. NEC Section 680.23(A)(5).

11-13 The smallest bonding conductor permitted for pool equipment is:

A) No. 14 AWG
C) No. 8 AWG
B) No. 6 AWG
D) No. 12 AWG

Answer: C

All metallic parts of the pool structure, including the reinforcing metal of the pool shell, coping stones, and deck, must be connected to a common bonding grid with a solid copper conductor, insulated, covered, or bare, not smaller than No. 8 AWG. NEC Section 680.26(C).

11-14 The smallest permitted equipment grounding conductor from a junction box to a panelboard supplying poolside equipment is:

A) No. 12 AWG
B) No. 14 AWG
C) No. 8 AWG
D) No. 6 AWG

Answer: A

A panelboard, not part of the service equipment, shall have an equipment grounding conductor installed between its grounding terminal and the grounding terminal of the service equipment. This conductor shall be sized in accordance with NEC Table 250.122 but not smaller than No. 12 AWG in any situation. NEC Section 680.25(B)(1).

11-15 The maximum voltage between conductors for lighting fixtures in pools and fountains is:

A) 120 volts
B) 150 volts
C) 300 volts
D) 480 volts

Answer: B

No underwater lighting fixtures shall be installed for operation at over 150 volts between conductors. NEC Section 680.23(A)(4).

11-16 Branch circuits feeding neon tubing installations must not be rated in excess of:

A) 10 amperes
B) 20 amperes
C) 15 amperes
D) 30 amperes

Answer: D

NEC Section 600.5(B)(2) requires branch circuits that supply neon tubing installations not exceed 30 amperes.

11-17 Which of the following connecting cables is permitted for use in connecting data processing systems to a branch circuit?

A) Computer/data processing cable and attachment plug cap

B) Small appliance connection cord

C) Flexible cord 25 feet long with an attachment plug cap

D) Cord-set assembly

Answer: D

NEC Section 645.5(B) permits cord-set assemblies for this use. The NEC now requires flexible cords with attachment plugs be used for this purpose. Such cords need to be 15 feet or less. It has also omitted computer/data processing cables and their associated plug caps as an acceptable wiring method. Small appliance cords are also not permitted for this use.

11-18 Power cables installed under raised floors in computer rooms are:

A) Not subject to requirements of the *NEC*

B) Not allowed

C) Not required to be secured in place

D) Allowed *only* if secured in place

Answer: C

NEC Section 645.5(E) does not require power cables, communications cables, connecting cables, interconnecting cables, and the like to be secured in place.

11-19 Electronic computer/data processing equipment shall be grounded in accordance with *NEC* Article 250 or:

A) Double insulated

B) Triple insulated

C) The grounding prong cut off of all plugs

D) Not used at all

Answer: A

Double insulated housings are considered to be adequate protection to guard against electrical shocks. NEC Section 645.15.

11-20 Each unit of an electronic computer supplied by a branch circuit must be supplied with a manufacturer's nameplate with the following information:

A) Name and address

B) Voltage, frequency, and maximum rated load in amperes

C) Phone number for service

D) Required conductor size and cord assembly information

Answer: B

A description of the required nameplate data is listed in NEC Section 645.16 under Marking.

11-21 The source of power for electrically-operated pipe organs must be:

A) A capacitor-start motor

B) From storage batteries only

C) A 4500-watt generator

D) A transformer-type rectifier

Answer: D

NEC Section 650.4 specifies the source of energy to be a transformer-type rectifier.

11-22 What is the maximum dc voltage allowed for the transformer-type rectifier in Question 11-21?

A) 25 volts

B) 30 volts

C) 125 volts

D) 240 volts

Answer: B

NEC Section 650.4 also limits the dc potential to 30 volts.

11-23 What *NEC* rating is given a piece of X-ray equipment with a rating based on an operating interval of 5 minutes or longer?

A) Long-time rating

B) Medium-time rating

C) Short-time rating

D) Miniature-time rating sometimes called ultra-time rating

Answer: A

This rating is defined under NEC Section 660.2, Definitions.

11-24 Branch-circuit conductors and overcurrent protective devices shall have at least the following percentage of x-ray equipment's momentary rating:

A) 50%

B) 100%

C) 120%

D) 150%

Answer: A

See explanation to Question 11-25.

11-25 Branch-circuit conductors and overcurrent protective devices for x-ray equipment shall have at least the following percentage of the equipment's long-time rating in all cases, except where otherwise specified in the *NEC*:

A) 50%

B) 100%

C) 125%

D) 150%

Answer: B

The ampacity of supply branch-circuit conductors and the overcurrent protective devices shall not be less than 50 percent of the momentary rating or 100 percent of the long-time rating, whichever is greater. NEC Section 660.6(A).

11-26 What is the minimum size fixture wire permitted for the control and operating circuits of X-ray and auxiliary equipment?

A) No. 12

B) No. 14

C) No. 18

D) No. 22

Answer: C

Size No. 18 or 16 fixture wires as specified in NEC Section 725.27 and flexible cords shall be permitted for the control and operating circuits of X-ray and auxiliary equipment where protected by not larger than 20-ampere overcurrent devices. NEC Section 660.9.

11-27 Which one of the following locations generally prohibits the use of induction heating equipment?

A) Dwellings

B) Industrial structures

C) Commercial buildings

D) Hazardous locations

Answer: D

Induction and dielectric heating equipment are not permitted in hazardous (classified) locations as defined in NEC Article 500 unless the equipment and wiring are designed for such use. NEC Section 665.4.

11-28 An assembly of electrically interconnected electrolytic cells supplied by a source of direct-current power is known as:

A) Cell line

B) Intercell

C) Cellular circuit

D) Electrolytic cellular communication

Answer: A

NEC Section 668.2 defines this assembly as a cell line.

Chapter 12

Transformers and Capacitors

Transformers play an important role in the distribution of electricity. Power transformers are located at generating stations to step up the voltage for more economical transmission. Substations with additional power transformers and distribution transformers are installed along the transmission line. Finally, distribution transformers are used to step down the voltage to a level suitable for utilization.

Transformers are also used quite extensively in all types of security/fire-alarm systems and heating/air-conditioning controls, or to raise and lower ac/dc voltages. It is important for anyone working with electricity to become familiar with transformer operations; that is, how they work, how they are connected into circuits, their practical applications and precautions to take while using them.

The chief use of capacitors is to improve the power factor of an electrical installation or an individual piece of electrically-operated equipment.

Since capacitors may store an electrical charge and hold a voltage that is present even when a capacitor is disconnected from a circuit, capacitors must be enclosed, guarded, or located so that persons cannot accidentally contact the terminals. In most installations, capacitors are installed out of reach or are placed in an enclosure accessible only to qualified persons. The stored charge of a capacitor must be drained by a discharge circuit either permanently connected to the capacitor or automatically connected when the line voltage of the capacitor circuit is removed. The windings of a motor or a circuit consisting of resistors and reactors will serve to drain the capacitor charge.

Capacitor circuit conductors must have an ampacity of not less than 135% of the rated current of the capacitor. This current is determined from the kVA rating of the capacitor as for any load. A 100 kVA (100,000 watts) three-phase capacitor operating at 480 volts has a rated current of:

100,000 volt-amperes/(1.73 × 480 volts) = 120.4 amperes

The minimum conductor ampacity is then:

1.35 × 120.4 amperes = 162.5 amperes

When a capacitor is switched into a circuit, a large inrush of current results to charge the capacitor to the circuit voltage. Therefore, an overcurrent protective device for the capacitor must be rated or set high enough to allow the capacitor to charge. Although the exact setting is not specified in the *NEC*, typical settings vary between 150% and 250% of the rated capacitor current.

In addition to overcurrent protection, a capacitor must have a disconnecting means rated at not less than 135% of the rated current of the capacitor unless the capacitor is connected to the load side of a motor-running overcurrent device. In this case, the motor disconnecting means would serve to disconnect the capacitor and the motor.

A capacitor connected to a motor circuit serves to increase the power factor and reduce the total kVA required by the motor-capacitor circuit. The power factor is defined as the true power in kilowatts divided by the total kVA, or:

$$pf = kW/kVA$$

where the power factor is a number between 0.0 and 1.0. A power factor less than 1.0 represents a lagging current for motors and inductive devices. The capacitor introduces a leading current that reduces the total kVA and raises the power factor to a value closer to unity. If the inductive load of a motor is completely balanced by the capacitor, a maximum power factor of unity results and all of the input energy serves to perform useful work; none is wasted.

The capacitor circuit conductors for a power factor correction capacitor must have an ampacity of not less than 135% of the rated current of the capacitor. In addition, the ampacity must not be less than one-third the ampacity of the motor circuit conductors.

The connection of a capacitor reduces current in the feeder up to the point of connection. If the capacitor is connected on the load side of the motor-running overcurrent device, the current through this device is reduced and its rating must be based on the actual current, not on the full-load current.

12-1 In cases where an oil-insulated transformer installation presents a fire hazard, which of the following safeguards must be used according to the degree of hazard?

A) Space separations

B) Fire-resistant barriers

C) Automatic fire suppression systems

D) Any of the above

Answer: D

NEC Section 450.27 allows any of these methods. This has been changed from an explanatory statement to a mandatory requirement.

12-2 Dry-type transformers rated over 35,000 volts must be installed:

A) Underground

B) At least 50 feet above floor level

C) In a vault

D) Pole-mounted only, with an adequate working platform

Answer: C

NEC Section 450.21(C) requires that all dry-type transformers rated over 35,000 volts be installed in a vault complying with Part III of NEC Article 450.

12-3 Potential transformers installed indoors require:

A) No overcurrent protection

B) Primary fuses

C) Secondary fuses only

D) A vault

Answer: B

NEC Section 450.3(C) requires that potential transformers installed indoors or enclosed must be protected with primary fuses.

12-4 Capacitor circuit conductors must have an ampacity at least what percent of the rated current of the capacitor?

A) 135%
B) 125%
C) 115%
D) 80%

Answer: A

NEC Section 460.8(A) gives the percentage as 135%. When connected to a motor terminal, the ampacity must not be less than one-third the ampacity of the motor circuit conductors and in no case less than 135% of the rated current of the capacitor.

12-5 A capacitor operating at 300 volts must be discharged to what voltage within 1 minute after it is disconnected from its supply?

A) 230 volts
B) 120 volts or less
C) 50 volts or less
D) 24 volts or less

Answer: C

The residual voltage of a capacitor must be reduced to 50 volts, nominal, or less, within 1 minute after the capacitor is disconnected from the source of supply. NEC Section 460.6(A).

12-6 Dry-type transformers installed indoors and rated 112½ kVA or less must have a minimum separation from combustible material of:

A) 6 inches
B) 8 inches
C) 10 inches
D) 12 inches

Answer: D

NEC Section 450.21(A) gives the minimum distance of 12 inches except when the transformer is separated from the combustible material by a fire-resistant heat-insulating barrier.

12-7 Unless specified otherwise in the *NEC*, the term "fire resistant" means a construction having a minimum fire rating of:

A) 1/2 hour

B) 1 hour

C) 2 hours

D) 3 hours

Answer: B

This is the time period given in NEC Section 450.21(B).

12-8 Dry-type transformers installed outdoors must have a:

A) Raintight enclosure

B) Weatherproof enclosure

C) Waterproof enclosure

D) Liquidtight enclosure

Answer: B

NEC Section 450.22 states "weatherproof" as opposed to the other terms. See NEC Article 100 for definitions of each.

12-9 What must be provided in each ungrounded conductor for each capacitor bank?

A) A disconnect and overcurrent protection

B) A disconnect only

C) Overcurrent protection only

D) None of these answers

Answer: A

NEC Sections 460.8(B) and 460.8(C) call for both; that is, a disconnecting means and overcurrent protection.

12-10 Each transformer over 600 volts must be protected by an individual overcurrent device on the primary side. Where fuses are used, the continuous current rating must not exceed what percent of the rated primary current of the transformer in an unsupervised location?

A) 115%

B) 125%

C) 150%

D) 300%

Answer: D

NEC Section 450.3(A) limits the fuse to 300% in an unsupervised location. Where the required fuse setting does not correspond to a standard rating, the next higher standard rating shall be permitted.

12-11 Name the three parts of a very basic transformer.

A) Fuses, conductors, and housing

B) Core, primary, and secondary windings

C) Three shunt windings

D) Core, taps, and one winding

Answer: B

See Figure 12-1.

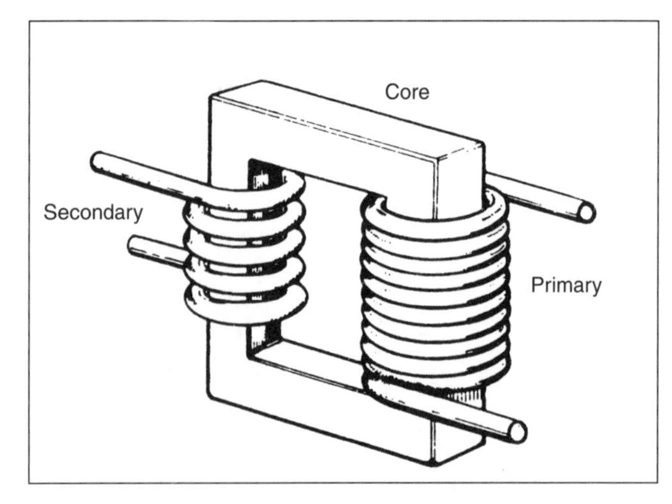

Figure 12-1: Components of a basic transformer

12-12 An autotransformer must have a continuous neutral current rating, in a 4-wire system, sufficient to handle:

A) The minimum neutral unbalanced load

B) The minimum neutral balanced load

C) The maximum possible neutral unbalanced load

D) The maximum neutral balanced load in the system

Answer: C

NEC Section 450.5(A)(4) requires the continuous neutral rating to handle the maximum possible neutral unbalanced load current of the 4-wire system.

12-13 The *NEC* requires that a fault sensing system be installed that will cause the opening of a main switch or common-trip overcurrent device for a three-phase, 4-wire autotransformer system. The sensing system is to guard against:

A) Internal faults only

B) Single-phasing and internal faults

C) Single-phasing only

D) Neither of these

Answer: B

NEC Section 450.5(A)(3) states that the system is to guard against both single-phasing and internal faults.

12-14 Name the three basic types of iron core transformers.

A) Open, closed, and shell

B) Auto, open, and closed

C) Encased, housed, and magnetic

D) Single-, two-, and three-phase

Answer: A

See Figure 12-2.

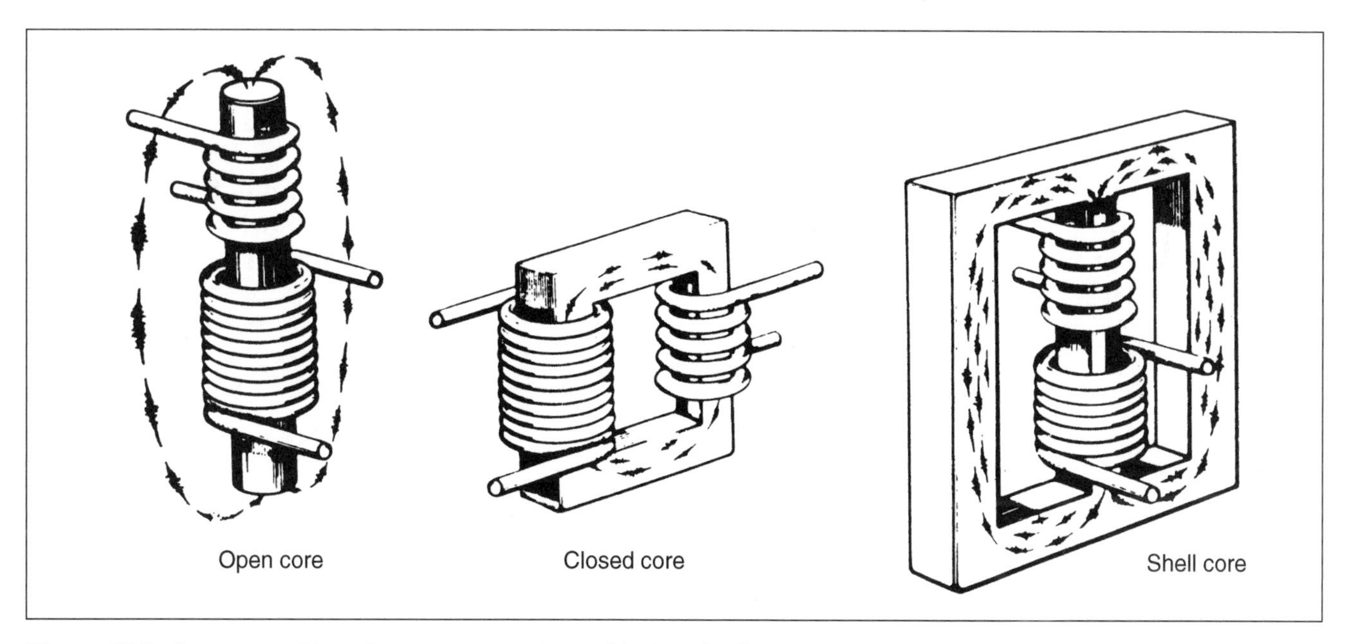

Figure 12-2: Iron cores of transformers are constructed in three basic types

12-15 What is the main purpose of transformer taps?

A) Adjust the wattage

B) Adjust the voltage

C) Change from step-up to step-down mode

D) Adjust the power-factor

Answer: B

Since voltage fluctuates as the distance from the generator increases, taps are installed on transformers to compensate for the variation in voltage.

12-16 Where fuses are used on the primary side of a transformer rated at over 600 volts, the fuse maximum current rating must not exceed what percent of the rated primary current of the transformer?

A) 125%

B) 150%

C) 250%

D) 300%

Answer: D

NEC Table 450.3(A) gives the rating as 300%.

12-17 Each autotransformer, 600 volts nominal or less, shall be protected by an individual overcurrent device installed in series with each ungrounded input conductor. Such devices shall be rated or set at not more than what percent of the rated full-load input current of the autotransformer?

A) 125%

B) 150%

C) 200%

D) 250%

Answer: A

NEC Section 450.4(A) states that such overcurrent devices must be rated or set at not more than 125% of the rated full-load input current of the transformer.

12-18 A fault-sensing system that will cause the opening of a main switch or common-trip overcurrent device for a three-phase, 4-wire transformer system must be provided to guard against single-phasing or internal faults. Name one method that will accomplish this requirement.

A) Using two subtractive-connected donut-type current transformers installed to sense and signal when an unbalance of 50% or more of rated current occurs in the line

B) Using three ammeters connected in series and one voltmeter connected in parallel

C) Using four voltmeters connected in parallel and one ammeter connected in series

D) Using a power-factor meter in conjunction with a watt-hour meter

Answer: A

NEC Section 450.5(A)(3)(FPN) gives the setup in Answer A as the correct method for establishing a fault-sensing system.

12-19 When a capacitor contains a certain amount of flammable liquid, it must be enclosed in a vault or an outdoor fenced enclosure. What amount of flammable liquid qualifies a capacitor for this condition?

A) Over 2 gallons

B) Over 3 gallons

C) Over 4 gallons

D) Over 5 gallons

Answer: B

Capacitors containing more than 3 gallons of flammable liquid shall be enclosed in vaults or outdoor fenced enclosures complying with NEC Article 110. NEC Section 460.2(A).

12-20 The rating or setting of an overcurrent device used on each ungrounded conductor for each capacitor bank 600 volts or less shall be rated:

A) As high as practicable

B) As low as practicable

C) The same as motor overload protection

D) Over 300%

Answer: B

NEC Section 460.8(B) states that the setting should be as low as practicable.

12-21 A thermal barrier must be provided if the space between resistors and reactors and any combustible material is less than:

A) 12 inches

B) 16 inches

C) 18 inches

D) 24 inches

Answer: A

NEC Section 470.3 calls for less than 12 inches.

12-22 How must resistors and reactors rated over 600 volts, nominal, be installed to protect personnel from accidental contact with energized parts?

A) Use overcurrent protection

B) Turn off current when personnel are in the area

C) Use warning signs

D) Isolate or elevate them

Answer: D

NEC Section 470.18(B) calls for resistors and reactors to be isolated by enclosure or elevation.

12-23 If there is no overcurrent protection on the secondary side, the rating of the primary overcurrent protective device for a 480-volt transformer with a rated primary current of 100 amperes should not exceed:

A) 100 amperes

B) 250 amperes

C) 125 amperes

D) 200 amperes

Answer: C

Each transformer 600 volts, nominal, or less, shall be protected by an individual overcurrent device on the primary side, rated or set at not more than 125% of the rated primary current of the transformer. Thus, 100 × 1.25 = 125 amperes. NEC Table 450.3(B).

12-24 Indoor installations of dry-type transformers greater than 112.5 kVA must be in an approved transformer room unless:

A) The transformer is rated with Class 155 insulation or higher and separated from combustibles with appropriate barriers

B) The transformer is rated at 75°C or lower

C) The transformer is rated at more than 35,000 volts

D) The transformer's primary is 2100 volts or less

Answer: A

The transformer must be rated at Class 155 insulation or higher and be separated from combustibles with appropriate barriers or by distances of 6 feet horizontally and 12 feet vertically. NEC Section 450.21(B). See Figure 12-3.

12-25 Dry-type transformers must be readily accessible except for:

A) Those rated at 601 volts or more and located in the open on walls, columns, or structures

B) Those with a full-load current rating of 250 amperes or less

C) Those with a full-load current rating of 125 amperes or less

D) Those rated at 600 volts or less and located in the open on walls, columns, or structures

Answer: D

NEC Section 450.13(A) does not require dry-type transformers to be readily accessible if the conditions in Answer D are met. See Figure 12-3.

12-26 Which of the following dry-type transformers must be installed in a vault when the transformer is located inside a building?

A) Transformers rated over 15,000 volts

B) Transformers rated over 25,000 volts

C) Transformers rated over 35,000 volts

D) Any transformer installed indoors

Answer: C

NEC Section 450.21(C) requires all indoor transformers rated at 35,000 volts or more to be installed in a transformer vault. See Figure 12-4 on the next page.

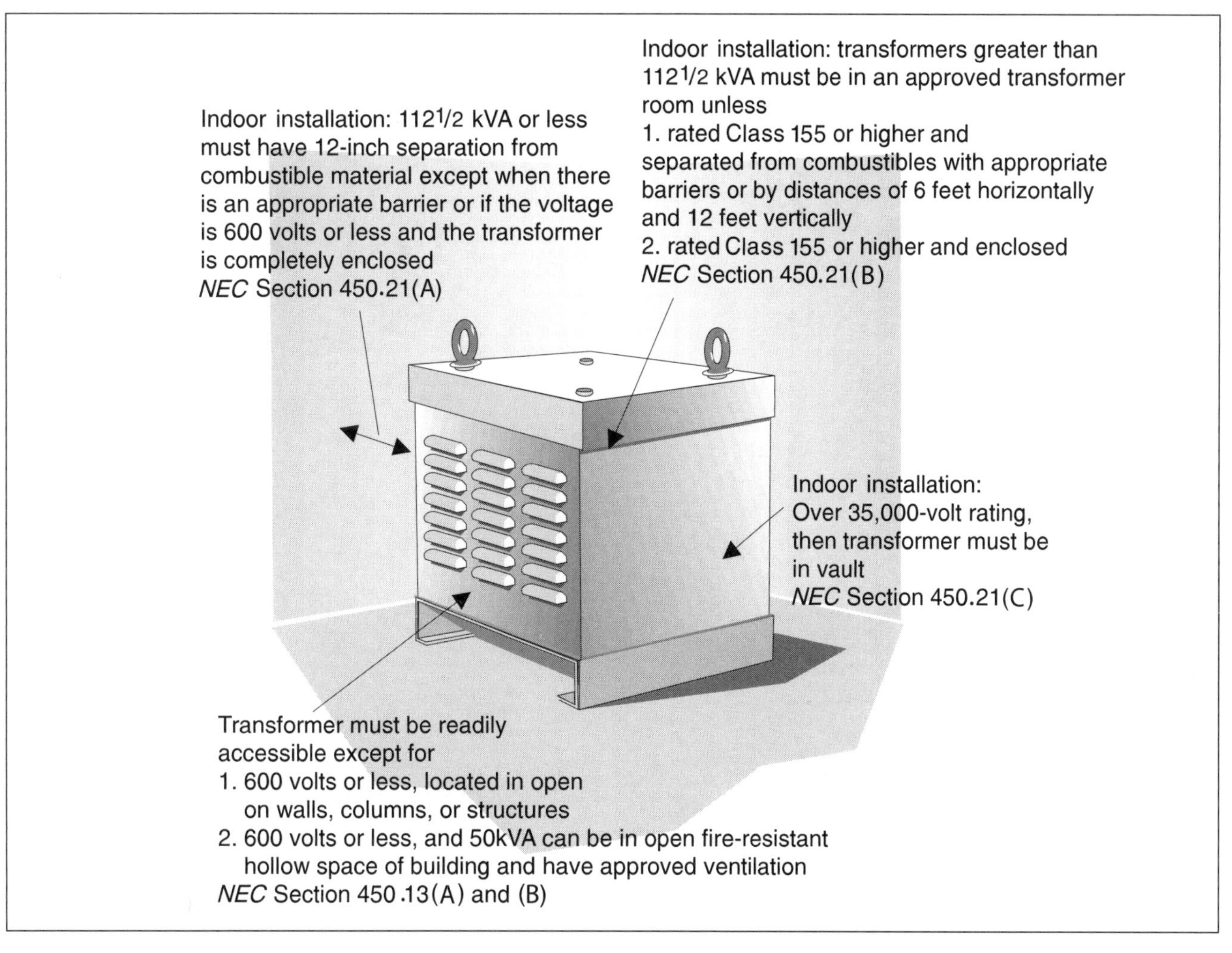

Figure 12-3: *NEC* installation requirements for dry-type transformers

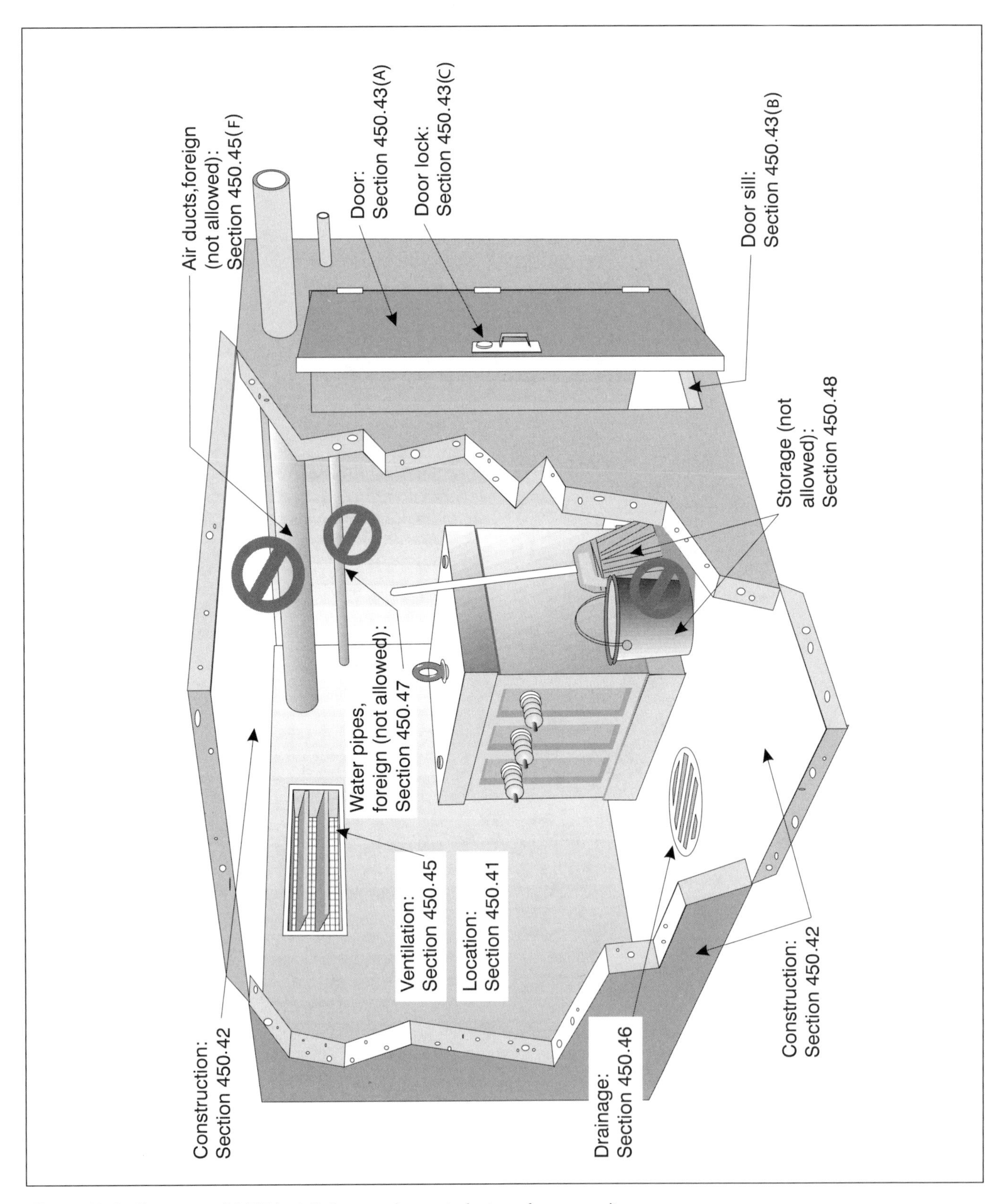

Figure 12-4: Summary of *NEC* installation requirements for transformer vaults

Chapter 13

Electrical Drawings

An electrical working drawing consists of lines, symbols, dimensions, and notations to accurately convey an engineer's design to the workers who install the electrical system on the job. Electrical drawings, therefore, use an abbreviated language for conveying a large amount of exact, detailed information, which would otherwise take many pages of manuscript or hours of verbal instruction to convey.

Every electrician, in every branch of electrical work, will need to consult and understand the information on electrical drawings to locate the various outlets, the routing of circuits, the location and size of panelboards, and other similar electrical details. The estimator of an electrical contracting firm must refer to electrical drawings to determine the quantity of material needed in preparing his bid for the work. Electricians in industrial plants consult schematic diagrams when wiring electrical controls for machinery. Plant maintenance personnel use electrical drawings in troubleshooting problems and to help locate faulty points in the installation.

Therefore, anyone contemplating getting their electrician's or contractor's license must have a good knowledge of electrical drawings. You must be able to read floor plans, schematic diagrams, and know the symbols used.

A complete set of working drawings for the average electrical system in industrial, commercial, and large residential projects will usually consist of the following:

- A plot plan showing the location of the building on the property and all outside electrical wiring, including the service entrance. This plan is drawn to scale with the exception of the various electrical symbols, which must be enlarged to be readable.

- Floor plans showing the walls and partitions for each floor level. The physical locations of all wiring and outlets are shown for lighting, power, signal and communication, special electrical systems, and related electrical equipment. Again, the building partitions are drawn to scale, as are such electrical items as fluorescent lighting fixtures, panelboards, and switchgear. The locations of other electrical outlets and similar components are only approximated on the drawings because they have to be exaggerated to be readable.

- Power-riser diagrams to show the service-entrance and panelboard components.

- Control wiring schematic and single-line diagrams.
- Schedules, notes, and large-scale details on construction drawings.

To be able to "read" electrical as well as other types of drawings, one must become familiar with the meaning of symbols, lines, and abbreviations used on the drawings and learn how to interpret the message conveyed by the drawings.

Electrical Specifications

Electrical specifications go hand in hand with electrical drawings in giving a written description of the work and the duties required of the owner, architect, and engineer. Together with the working drawings, the written specifications form the basis of the contract requirements for the construction of a building's electrical system.

The questions in this chapter are designed to review electrical drawings and written specifications to give you some idea of the type of questions that might appear on electrician's exams.

In reviewing these questions, please be aware that electrical symbols may vary on different drawings, but in actual practice there is usually a symbol list or legend giving the exact meaning of each. It is recommended that you review several books on the subject, as well as actual working drawings, if you find that blueprint reading is one of your weak areas.

Division 16 — ELECTRICAL

Section 16A — General Provisions

1. The "Instructions to Bidders," "General Conditions," and "General Requirements" of the architectural specifications govern work under this section.

2. It is understood and agreed that the Electrical Contractor has, by careful examination of the Plans and Specifications, and the site where appropriate, satisfied himself as to the nature and location of the work, and all conditions which must be met in order to carry out the work under this Section of the Contract.

3. **The Scope of Work**

a. The scope of work consists of the furnishing and installing of complete electrical systems — exterior and interior — including miscellaneous systems. The Electrical Contractor shall provide all supervision, labor, materials, equipment, machinery, and any and all other items necessary to complete the systems. The Electrical Contractor shall note that all items of equipment are specified in the singular; however, the Contractor shall provide and install the number of items of equipment as indicated on the drawings and as required for a complete system.

Sample: Excerpt from a typical electrical specification for an actual building project

13-1 A "section" or "cross-section" of an object or a building is what could be seen if the object were:

A) Sliced into two parts with one part removed

B) Sliced into six parts with three parts removed

C) Sawed into four parts

D) Left solid

Answer: A

A section of an object is what could be seen if the object were cut or sliced into two parts at the point where the section is taken; then the portion between the viewer and the cutting plane is removed to reveal the interior details of the object.

13-2 A supplemental drawing used with conventional electrical drawings that gives a complete and more exact description of an item's use is called a:

A) Title block

B) Detail drawing

C) Schedule

D) Riser diagram

Answer: B

An electrical detail drawing is a drawing of a single item or a portion of an electrical system. It gives all the necessary details and a complete description of its use to show workers exactly what is required for its installation.

13-3 A site plan of a building is a plan view (as if viewed from an airplane) that shows:

A) Each floor level of the building

B) Power-riser diagrams

C) Cross-sections of the building

D) Property boundaries and buildings

Answer: D

A site plan shows the property boundaries and the building(s) drawn to scale and in its (their) proper location on the lot. Such plans will also include sidewalks, drives, streets, and similar details. Utilities such as water lines, sanitary sewer lines, telephone lines, and electrical power lines also appear on site plans.

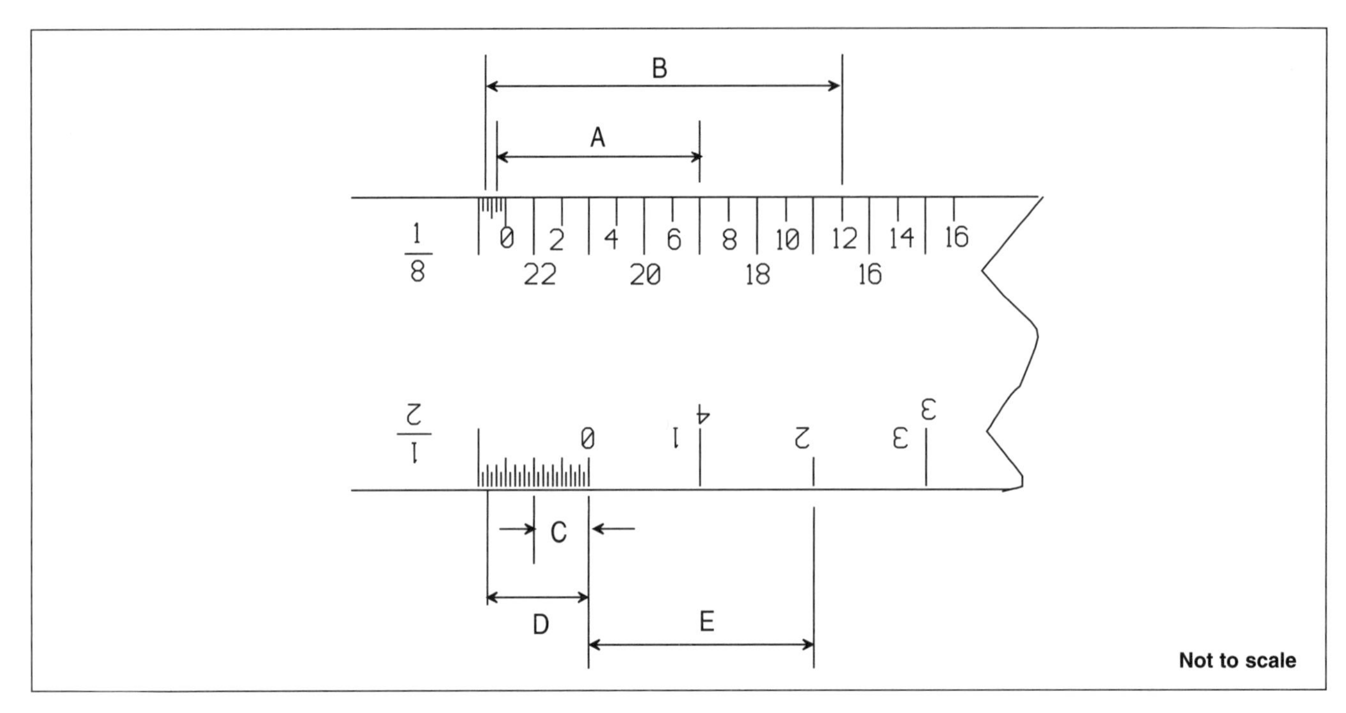

Figure 13-1: The 1/8 inch architect's scale used to measure building dimensions

13-4 Using the 1/8 inch architect's scale in Figure 13-1, what is dimension "A"?

A) 12 feet 10 inches

B) 7 feet 4 inches

C) 12 feet 6 inches

D) 11 feet

Answer: B

Reading the 1/8 inch scale from left to right, the inch scale shows a dimension of 4 inches. Continue from the zero mark on the foot scale to the right to 7; thus, the reading is 7 feet 4 inches.

13-5 Using the 1/8 inch architect's scale in Figure 13-1, what is dimension "B"?

A) 12 feet 10 inches

B) 7 feet 4 inches

C) 12 feet 6 inches

D) 11 feet

Answer: A

Reading the 1/8 inch scale from left to right, the inch scale shows a dimension of 10 inches. Continue from the zero mark on the foot scale to the right to 12; thus, the reading is 12 feet 10 inches.

13-6 Look at the architect's scale in Figure 13-1. What is dimension "C" on the 1/2 inch scale?

A) 6 inches

B) 8 inches

C) 10 inches

D) 14 inches

Answer: A

Each mark on the inch scale represents 1/2 inch. Since 12 marks are covered, this is 6 inches.

13-7 Look at the architect's scale in Figure 13-1. What is dimension "D" on the 1/2 inch scale?

A) 6 inches

B) 8 inches

C) 10 inches

D) 11 inches

Answer: D

Each mark on the inch scale represents 1/2 inch. Since 22 marks are covered, this is 11 inches.

13-8 Look at the architect's scale in Figure 13-1. What is dimension "E" on the 1/2 inch scale?

A) 1 foot

B) 2 feet

C) 3 feet

D) 4 feet

Answer: B

Each mark on the foot scale represents 1 foot. Since 2 marks are covered, this represents 2 feet.

13-9 What is dimension "A" on the 1/4 inch scale in Figure 13-2 on the next page?

A) 4 feet 9 inches

B) 1 foot 9 inches

C) 1 foot 6 inches

D) 37 feet 0 inches

Answer: A

Each mark on the inch scale represents 1 inch; each mark on the foot scale represents 2 feet. Consequently, the measurement of "A" is 4 feet 9 inches.

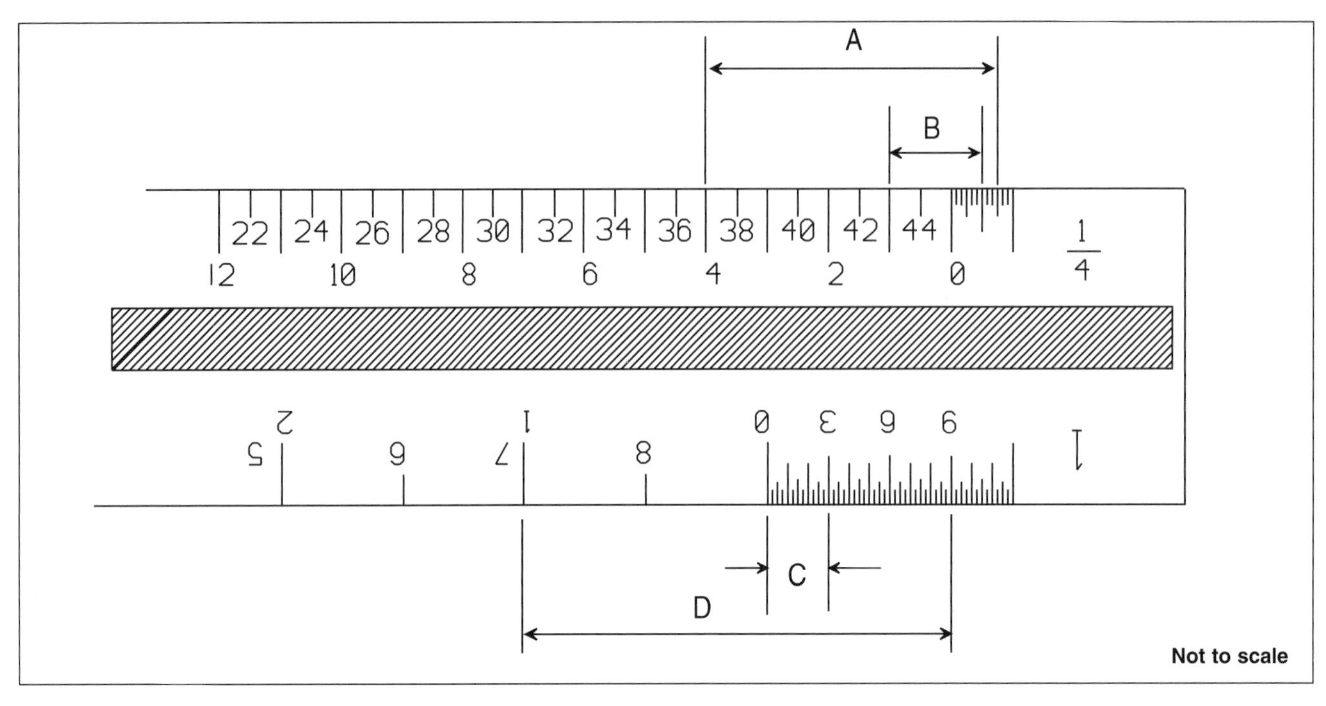

Figure 13-2: Typical architect's scale

13-10 What is dimension "B" on the 1/4 inch scale in Figure 13-2?

A) 1 foot 6 inches

B) 4 feet 6 inches

C) 43 feet 6 inches

D) 6 feet 4 inches

Answer: A

This scale is read from right to left; each mark in the inch section of the scale represents 1 inch. Therefore, 6 inches are shown in the inch section and 1 foot is shown in the foot section. Thus, 1 foot 6 inches.

13-11 What is dimension "C" on the 1 inch scale in Figure 13-2?

A) 6 inches

B) 3 inches

C) 2 inches

D) 1 inch

Answer: B

Each of the longer marks on the inch scale represents 1 inch. Since there are three long marks, the measurement is 3 inches.

13-12 What is dimension "D" on the 1 inch scale in Figure 13-2?

A) 1 foot 6 inches

B) 12 inches

C) 1 foot 9 inches

D) 12 feet

Answer: C

This inch section of the 1 inch scale is read from left to right; each long mark represents 1 inch; the reading is 9 inches. The foot scale is read from right to left, indicating 1 foot. Therefore, the measurement is 1 foot 9 inches.

13-13 At which of the following drawing locations are the receptacle layouts most likely to be?

A) Floor plan

B) Plot plan

C) Panelboard schedule

D) Site plan

Answer: A

The floor plan typically contains information such as receptacle layouts, GFCIs, etc.

13-14 At which of the following drawing locations is the drawing scale most likely to be?

A) In the symbol list

B) In the power-riser diagram

C) Panelboard schedule

D) Title block

Answer: D

The drawing scale is usually indicated in the drawing title block, but scale markings may also be found under other supplemental views found on the drawing sheet.

13-15 A legend or electrical symbol list is shown on electrical working drawings to:

A) Describe materials and installation methods

B) Show the outline of the architect's floor plan

C) Identify all symbols used to indicate electrical outlets or equipment

D) Enable the electric service size to be calculated

Answer: C

Electrical symbols vary, so a legend or electrical symbol list usually appears on drawings to show the meaning of each symbol. See Figure 13-3. Where a symbol is used to identify a special electrical component, sometimes a note is used adjacent to the symbol to describe it.

13-16 Which of the following best describes the symbol normally used for duplex receptacles?

A) A square box with the letter "R" next to it

B) A circle with one line drawn through it

C) A square box with an "X" drawn inside

D) A circle with two parallel lines drawn through it

Answer: D

The symbol described in Answer D is the one normally used. See Figure 13-3.

13-17 Of the sixteen usual divisions found in written specifications for a building project, which division deals mainly with electrical work?

A) Division 2

B) Division 4

C) Division 8

D) Division 16

Answer: D

Division 16 is the one dedicated to electrical work. Specifications for electrical and electronic controls for HVAC systems may also be found in Division 15.

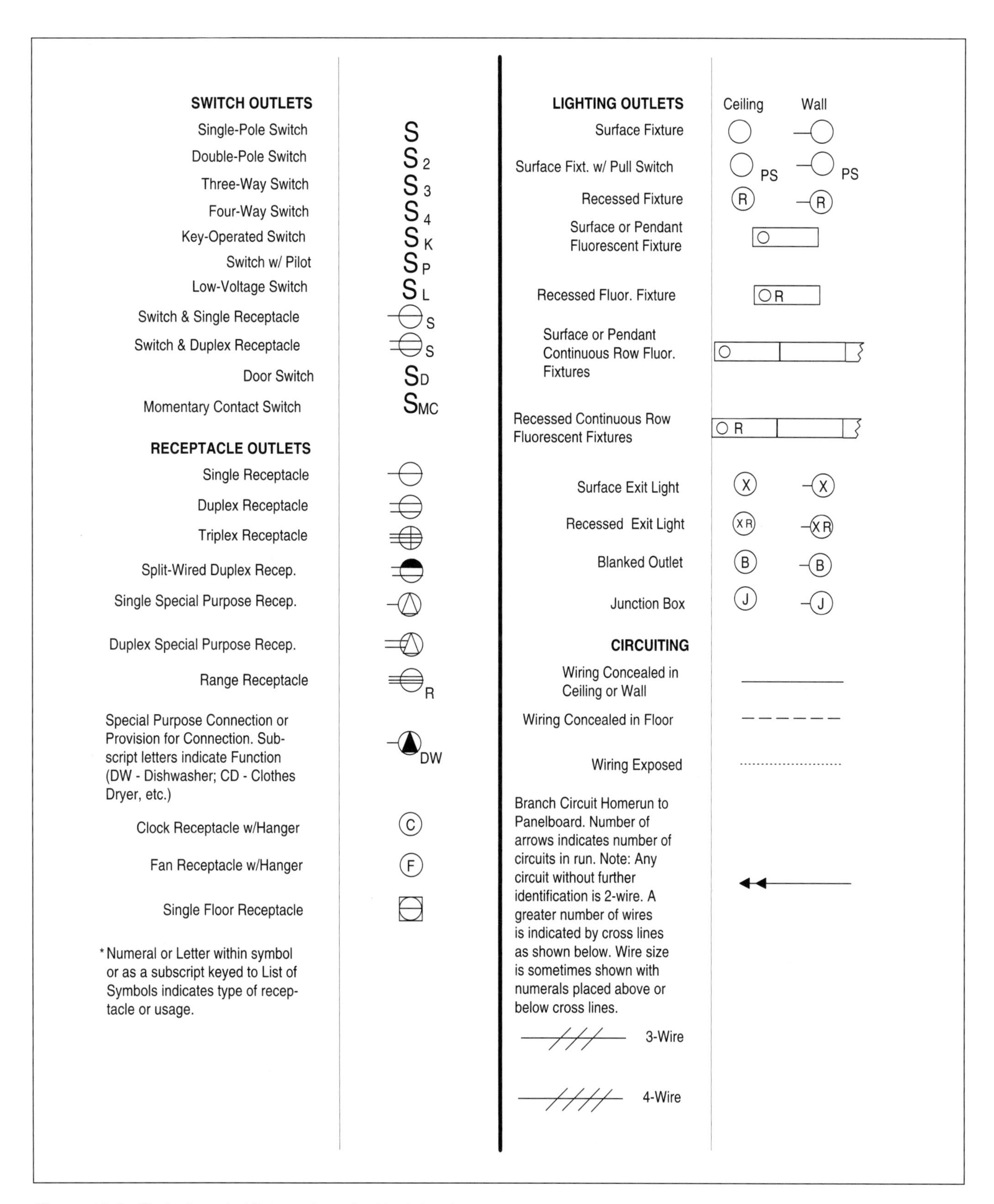

Figure 13-3: Typical symbol list used on electrical drawings

13-18 Look at the drawing in Figure 13-4. How many duplex receptacles are shown in the living room?

A) Two
B) Four
C) Six
D) Eight

Answer: C

There are six duplex receptacle symbols shown in the living room in this drawing.

13-19 How many branch circuits are shown in Figure 13-4?

A) Five
B) Six
C) Seven
D) Eight

Answer: C

Two circuits serve the kitchen, 1 serves the master bedroom, 1 serves both bedroom #2 and bedroom #3, 1 serves the living room/vestibule, and 1 supplies the GFCI circuit which supplies protection to the carport receptacle, both front and rear receptacles, NEC Section 210.8(A). One circuit supplies the GFCI receptacle in the bathroom, NEC Section 210.11(C)(3).

13-20 How many duplex receptacles are installed outside the building in the drawing in Figure 13-4?

A) One
B) Two
C) Three
D) Four

Answer: C

Three outside duplex receptacles are shown by symbol on the floor-plan drawing and also described in the Note.

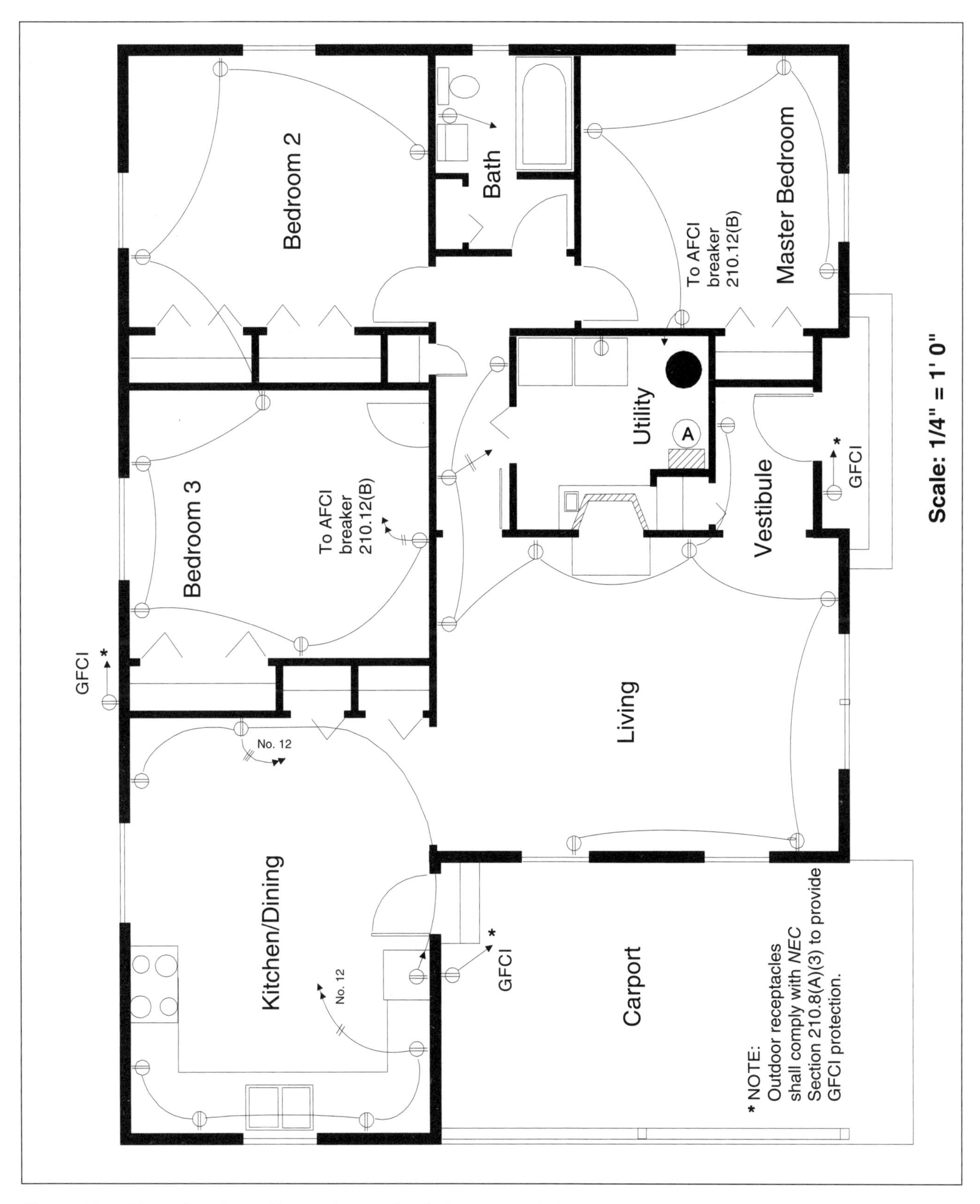

Figure 13-4: Floor plan of a residence showing the duplex receptacle layout

13-21 To comply with the *NEC*, what device must be used on all three outside duplex receptacles shown on the drawing in Figure 13-4?

A) Double-pole circuit breaker

B) A ground-fault protector (ground-fault circuit interrupter)

C) A single-pole mercury switch

D) A 40-amp circuit breaker

Answer: B

A ground-fault protector is indicated by notes on the drawing in Figure 13-4. See Section 210.8(A)(3) of the NEC.

13-22 To what scale is the floor plan of the residence in Figure 13-4 drawn?

A) 1/16" = 1'-0"

B) 1/8" = 1'-0"

C) 1/4" = 1'-0"

D) 1" = 1'-0"

Answer: C

Scale: 1/4" = 1'-0" is indicated by note on the drawing.

13-23 In the power-riser diagram in Figure 13-5, what does "C/T cab." stand for?

A) Central telephone cabinet

B) Current transformer cabinet

C) Control terminal cabinet

D) Central termination cabinet

Answer: B

The service conduit and conductors enter the C/T cabinet where current transformers are used in conjunction with the electric meter for metering the amount of power used.

13-24 What size and how many service conductors are installed in each of the 3 1/2 inch conduits from the service head to the C/T cabinet in Figure 13-5?

A) Four 3/0 conductors

B) Four No. 1 AWG conductors

C) Eight 350 kcmil conductors

D) Three No. 10 AWG conductors

Answer: C

Eight 350 kcmil conductors are noted on the drawing.

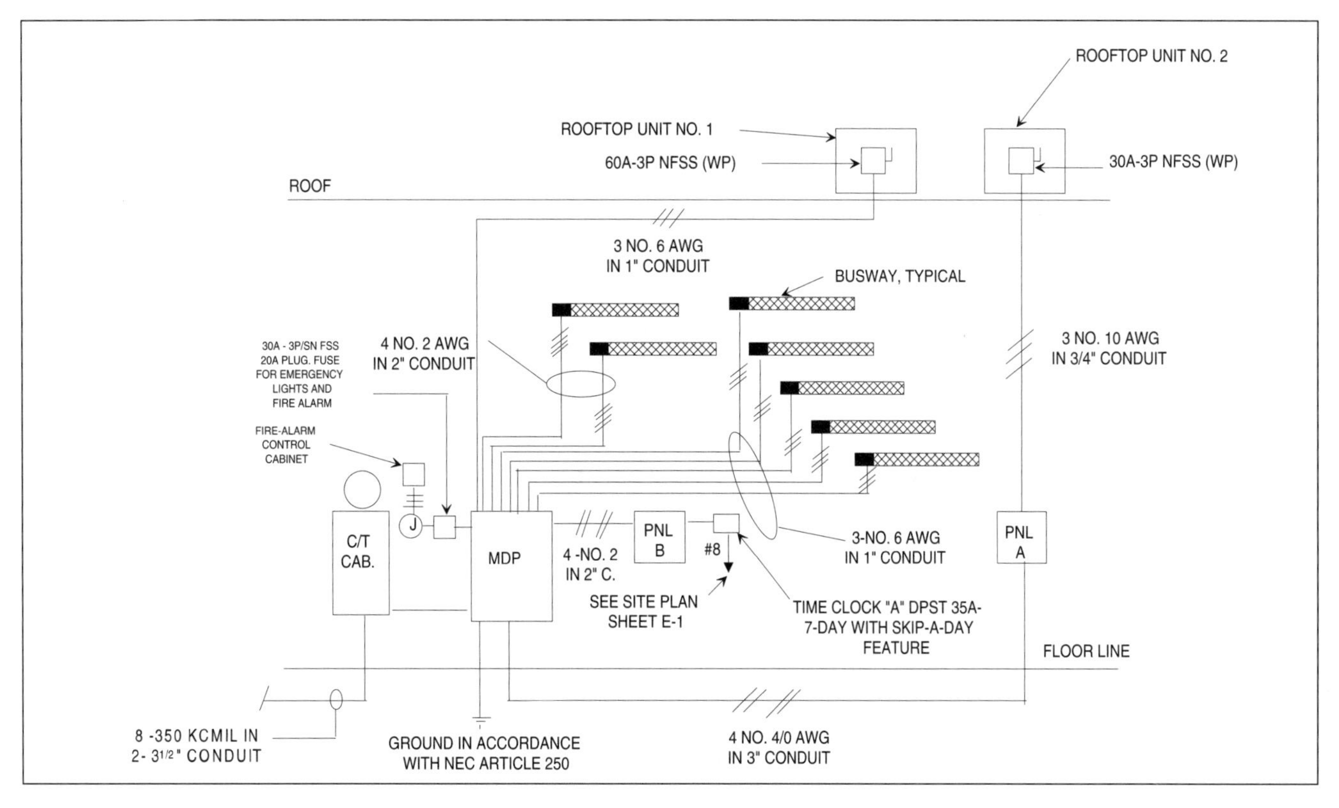

Figure 13-5: Typical power-riser diagram

13-25 What size conduit is shown in the drawing in Figure 13-5 for containing three No. 6 AWG conductors?

A) 1 inch

B) 2 inch

C) 3 inch

D) 4 inch

Answer: A

The notes on the drawing indicate that three No. 6 AWG conductors are to be installed in 1-inch conduit.

13-26 Which of the following best describes where the type and size of overcurrent protection may be found?

A) Floor plans

B) Lighting-fixture schedule

C) Panelboard schedule

D) Cross-sectional views

Answer: C

The panelboard schedule on drawings usually indicates the type and size of overcurrent protection.

13-27 What wire size and how many conductors are specified for the feeder circuit supplying the rooftop unit No. 2 in the drawing in Figure 13-5?

A) Three No. 1 AWG
B) Three No. 10 AWG
C) Four No. 2 AWG
D) Two No. 4 AWG

Answer: B

The note adjacent to the feeder indicates three No. 10 AWG conductors.

13-28 What size conduit is specified to contain the conductors in Question 13-27?

A) 2 inch
B) 1½ inch
C) 1 inch
D) ¾ inch

Answer: D

The same note as described in Question 13-27 also states that these conductors are to be installed in ¾-inch conduit.

13-29 How many sections of busway are specified in the drawing in Figure 13-5?

A) 7
B) 10
C) 17
D) 27

Answer: A

There are seven sections shown in the power-riser diagram in Figure 13-5.

13-30 What size wire is used to feed the two sections of bus duct on the left side of Figure 13-5?

A) No. 1 AWG
B) No. 2 AWG
C) No. 3 AWG
D) No. 4 AWG

Answer: B

A note adjacent to these two sections of bus duct states that a wire size of No. 2 AWG is to be used.

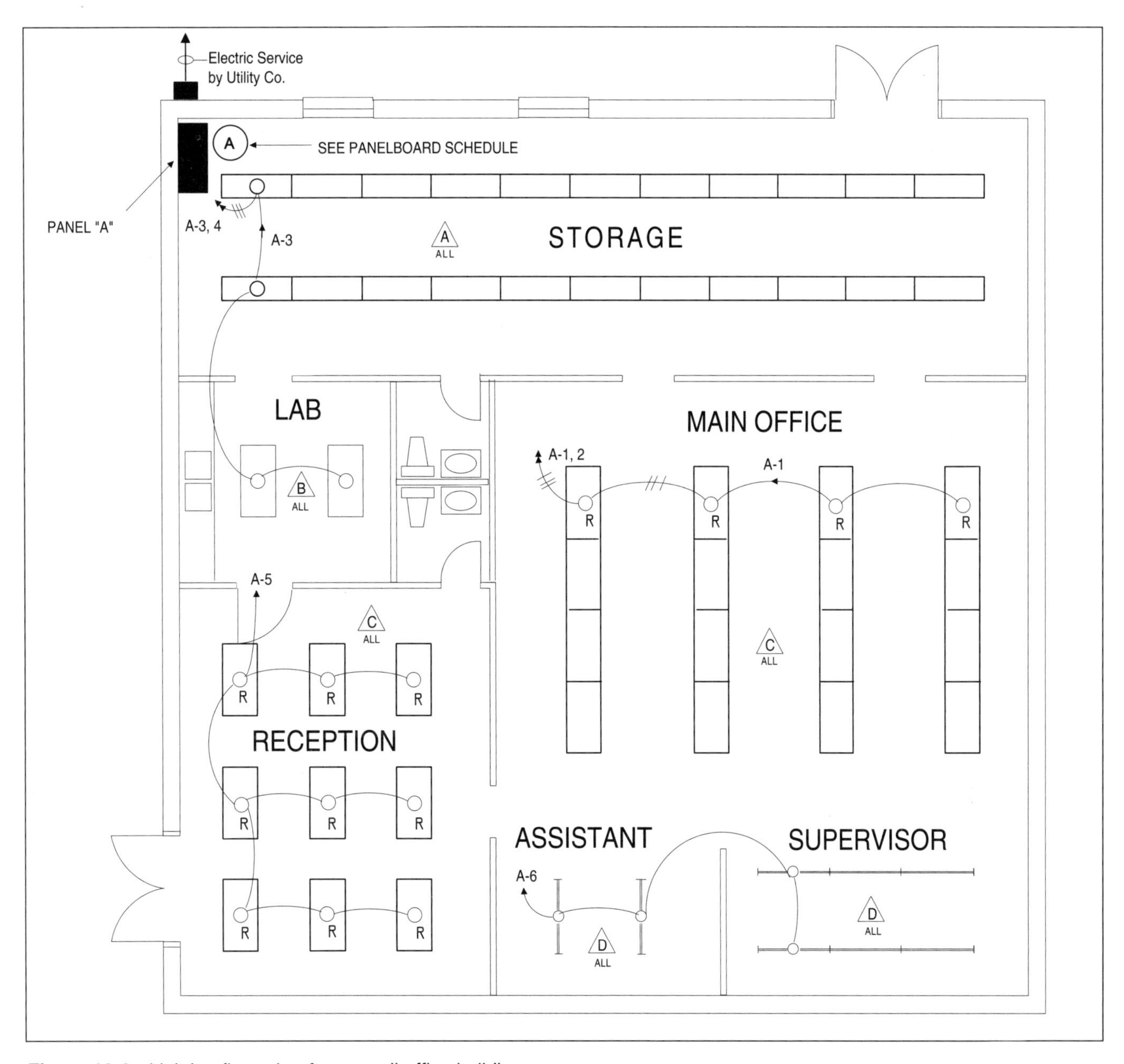

Figure 13-6: Lighting floor plan for a small office building

13-31 Who is to provide the electric service for the project in Figure 13-6?

A) Electrical contractor

B) Utility company

C) Electrical engineer

D) Architect

Answer: B

A note adjacent to the electric meter states that the utility company is to provide the electric service.

13-32 If each of the Type C lighting fixtures in Figure 13-6 have a total lamp and ballast load of 200 volt-amperes, what is the total connected load of circuit A-1 (in volt-amperes)?

A) 1600 volt-amperes

B) 1700 volt-amperes

C) 1800 volt-amperes

D) 1900 volt-amperes

Answer: A

Since there are eight fixtures fed by this circuit, 8 × 200 = 1600 volt-amperes.

13-33 How many of each type lighting fixture are connected to circuit A-3 in Figure 13-6?

A) 22 Type A fixtures

B) 8 Type D fixtures and 9 Type C fixtures

C) 16 Type C fixtures

D) 11 Type A fixtures and 2 Type B fixtures

Answer: D

The two Type B fixtures in the lab area are run to the junction box at the row of Type A fixtures; the circuit symbol A-3 then has a total of 11 Type A fixtures and 2 Type B fixtures connected to it.

13-34 Type C fixtures in Figure 13-6 have the letter "R" next to the outlet boxes. According to the ANSI electrical symbols in Figure 13-3, what does this "R" mean?

A) The lighting fixtures are raised to the surface of the finished ceiling

B) The lighting fixtures are recessed

C) The lenses are colored red

D) The lighting fixtures are resistant to insects

Answer: B

The letter "R" next to a lighting fixture indicates that the fixture is recessed.

13-35 Where does the drawing in Figure 13-6 indicate that information can be found concerning panel "A"?

A) In the lighting-fixture schedule

B) In the written specifications

C) In the panelboard schedule

D) In the kitchen equipment schedule

Answer: C

A note adjacent to panel "A" instructs those reading the drawings to see panelboard schedule.

13-36 How are fixture types identified in the lighting floor plan in Figure 13-6?

A) A letter inside a triangle

B) A numeral inside a triangle

C) A letter inside a square box

D) A numeral inside a square box

Answer: A

This drawing uses a letter inside a triangle to indicate the fixture type. Although not shown, a lighting-fixture schedule is normally used to describe each fixture type in more detail.

13-37 What is one good reason that circuit lines are drawn curved rather than straight?

A) This is how conduit is installed in buildings

B) To enable the drafters to route the lines around partitions

C) So as not to confuse the circuit lines with building lines

D) Curved lines are easier to draw on CAD systems than straight lines

Answer: C

When circuit lines are drawn straight, they are sometimes confused with the building lines. Thus, Answer C is one good reason for drawing curved circuit lines.

13-38 What do arrowheads placed on circuit lines mean?

A) Designates the room in which the circuit is installed

B) Homeruns to the designated panel

C) The direction of current flow

D) Designates that the circuit is to be controlled by a wall switch

Answer: B

Arrowheads are used to indicate a homerun to a panelboard.

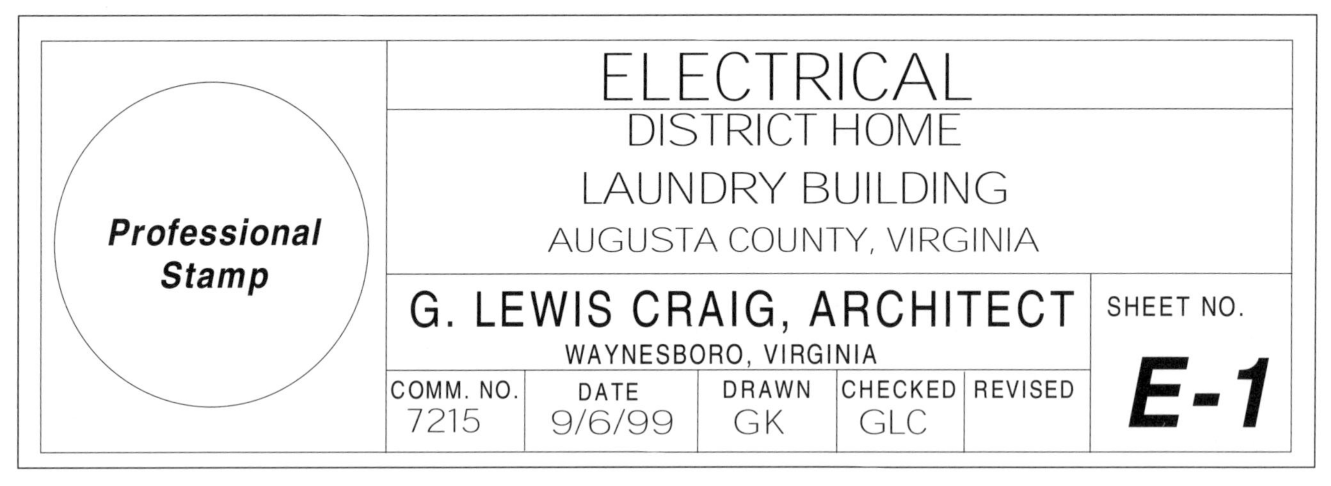

Figure 13-7: A typical title block for architectural drawings

13-39 The drawing title block in Figure 13-7 shows that the project is located in what county or city?

A) Interior County

B) Ruty, Maryland

C) Waynesboro, Virginia

D) Augusta County

Answer: D

Augusta County, Virginia is named in the space beneath the building name.

13-40 The drawing title block in Figure 13-7 shows that the drawing sheet in this set of drawings is:

A) E-1

B) GK

C) E-3

D) 7215

Answer: A

The Sheet No. block indicates this drawing sheet to be E-1. Sometimes this block will also indicate the total number of pages in the drawing set; that is, "E-1 of 2," etc.

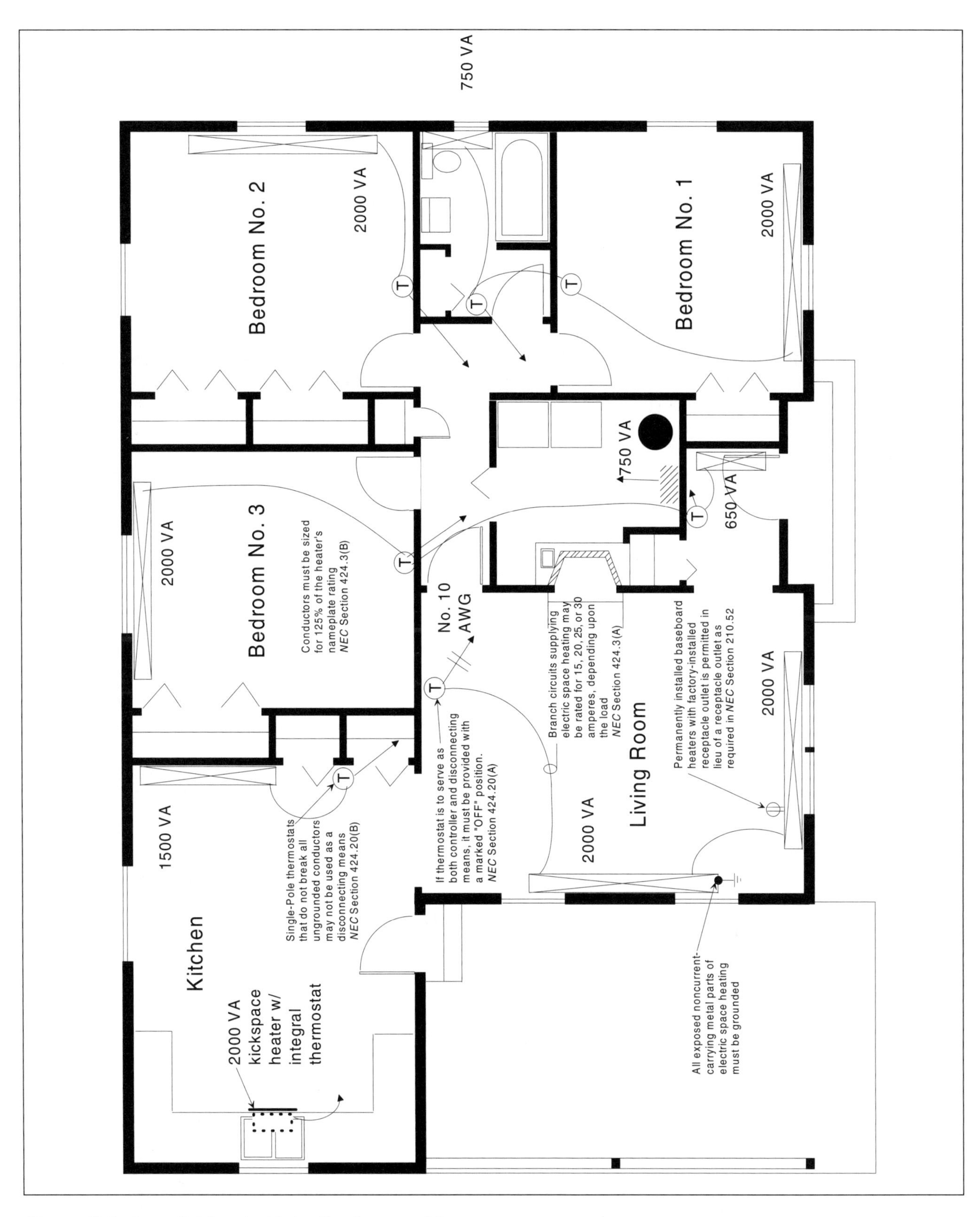

Figure 13-8: Layout of the electric heating for a small home

13-41 From the title block in Figure 13-7, what does this sheet of the drawing entail; that is, which of the trades?

A) HVAC
B) Electrical
C) Plumbing
D) Interior design

Answer: B

The main sheet title is "Electrical."

13-42 The drawing in Figure 13-8 shows the floor plan of a residence with a combination of electric baseboard and forced-air (down-flow) heaters in the various rooms. If the long rectangular symbols with an "X" in them represent the baseboard heaters, how many are to be installed in the residence?

A) 4
B) 6
C) 8
D) 10

Answer: C

There is one in the kitchen, one in each bedroom, one in the bath, one in the foyer, and two in the living room. The one in the utility room is a rectangle with hatch marks, signifying a wall-mounted heater.

13-43 The wall-mounted thermostats in Figure 13-8 are designated by a circle with the letter "T" inside. How many of these thermostats are shown?

A) 5
B) 6
C) 7
D) 8

Answer: C

There is one in the kitchen, one in each bedroom, one in the bath, one in the foyer, and one in the living room.

13-44 Where are all the wall-mounted thermostats in Figure 13-8 located?

A) On outside walls
B) On the floor
C) On inside walls
D) On the ceiling

Answer: C

The drawing shows all wall-mounted thermostats mounted on an inside wall (partition); that is, a wall that is not exposed to the outside air temperature.

Final Examination

Now that you have gone through all the questions in this book (and reviewed supplemental material in areas where you were weak), you should be ready for that big day — the exam to test your ability as an electrician or an electrical contractor. But you just don't walk into the examiner's office and say you want to take the exam; it's not quite that simple. First of all you have to get permission to take the examination which normally requires filling out an application and paying a fee. The examining board then reviews your application and either approves or disapproves it. If approved, you will then be notified of the time and place to take the actual examination.

Where to Apply

If you want to apply for your journeyman electrician's license, this is handled in several different ways in various parts of the country. For example, many industrial establishments and labor organizations prefer to train their own electricians through various types of training programs. This is usually a four- or five-year program consisting of on-the-job training along with some classroom instruction. At the end of this period, and after passing an examination, the apprentice or trainee is awarded a certificate.

In some localities, anyone who "runs" a job for a contractor, holding the position of supervisor, foreman, general foreman, or job superintendent, must also have a master electrician's license. These exams are normally given by the local electrical inspector's office. An electrical contractor's license is almost always required in any area if the person is contracting electrical work.

Other local electrician's exams are given by various agencies, but most will fall under the city or county electrical inspector's office. Look in your local phone directory under your town or city. For example, in Luray, Virginia you would look under "Luray — Town of." Then look for "Electrical Inspector." Should your town not have an electrical inspection department, contact the building inspector's office. This office should be able to direct you to the proper place to apply for your electrician's exam. If not, contact your local power or utility company.

More and more states now require that all electrical contractors must be licensed by the state. A state license usually permits the contractor to work in any city, town or county within the state. Appendix I of this book gives the names and addresses of state licensing agencies throughout the United States.

To apply for your state electrical contractor's license, contact the appropriate agency listed in the Appendix. In due time you will receive the various application forms. When applying for your contractor's license, keep in mind that you are not only tested on your knowledge of electrical installations

Figure 14-1: The electrician's examination is designed to test your knowledge of eletrical theory and *NEC* installation requirements

and theory, but also on your past experience and business ability.

The next several pages show actual forms that are required to be filled out for the State of Tennessee. Other states should be similar.

Instructions: The first page of the packet should be filled out with your name, complete address, telephone number and date of financial statement. Page 2 (Figure 14-2) contains general instructions on how to fill out the remaining portion of the form.

Application: Figure 14-3 shows the next 5 pages of the application which include other areas where you hold a license, list of references, your statement of experience, bank references, equipment suppliers, and the like. You will also have to provide a financial statement, affidavit, and other information about your past experience and business.

On Examination Day

The exact schedule for electrician's exams will vary from locale to locale. However, the following is typical of states giving an electrical contractor's exam:

- 7:30 a.m.: Examinee reports to the examination center; seats assigned.
- 8:00 a.m.: Examination orientation begins.
- 8:30 a.m.: Morning examination session begins.
- 12:30 p.m.: Morning examination session ends.
- 1:15 p.m.: Examinee reports to examination center.
- 1:30 p.m.: Afternoon examination session begins.
- 5:30 p.m.: Afternoon examination session ends.

To ensure that all examinees are examined under equally favorable conditions, the following regulations and procedures are observed at most places of examinations:

- Each examinee must present proper identification, preferably a driver's license, before being permitted to take the examination.
- Each examinee will be assigned to a specifically designated seat and this seat will remain your seat assignment for the entire examination. Once you enter the examination room, locate your assigned seat and be seated.
- Each examinee should bring a watch. No one will be permitted to work beyond the established time limits.
- Examinees should not, in most cases, bring books or other reference materials to the examination center unless instructed to do so in your state's application. Many states will not allow the examinee to take the exam if any notes or other reference material are brought to the exam center.
- In most states, examinees will be furnished with two #2 black lead pencils and a current copy of the *National Electrical Code* for use during the examination. Other states have no open book exams at all.

INSTRUCTIONS

1. Complete questionnaire form in its entirety and return to:
   Board for Licensing Contractors
   500 James Robertson Parkway
   Suite 110
   Nashville, TN 37243

2. Attach a check for $150.00 made payable to: Regulatory Boards/Contractors. **This fee is nonrefundable.**

3. Any questions you think do not apply to you, write in N/A. Do not leave any questions blank.

4. The financial statement must be reviewed or audited and prepared by a Certified Public Accountant or Licensed Public Accountant. A financial report prepared on their letterhead or in their booklet form is acceptable as an alternative to completing the pages provided in this questionnaire.

5. All corporations engaged in contracting in the state are required to register with the Tennessee Secretary of State and receive either a charter (DOMESTIC CORP.) or certificate of authority (FOREIGN CORP.). A copy of the charter or certificate must be attached. Contact the Secretary of State at (615) 741-2286.

6. All letters of reference **must** have the name and address of the person completing the form in the upper left hand corner.

7. Reference forms can be removed from the questionnaire and distributed but they must be returned with the completed questionnaire.

8. Submittal of a line of credit is **optional**. The line of credit will be used to increase your working capital **only**. It must be in the format as used in the sample on page 23.
   (100% allowed)

9. Submittal of a guaranty agreement is **optional**. The agreement must be accompanied by either your personal financial statement (partnerships and corporations) or the financial statement of another corporation. It must be signed and notarized. This will go toward increasing your net worth **only**. Refer to page 21.
   (by Board rule only 50% allowed)

10. The financial statement work sheet is to be completed by your CPA or Licensed PA. It is to be removed from the questionnaire and brought with you to your Board interview. The Board or staff will be unable to provide assistance in the preparation of the worksheet.

2

Figure 14-2: General instructions for electrical contractor's license

APPLICATION
STATE OF TENNESSEE
BOARD FOR LICENSING CONTRACTORS

Date ______________________

1. Check your manner of operation:
   (   ) Corporation   (   ) Partnership   (   ) Individual

2. List names and title(s) of individual, partners, or corporate officers:
   ____
   ____
   ____

3. State specifically each branch of contracting for which you are requesting to be licensed:
   ____
   ____

   State monetary limit desired ______________________

4. Have you ever held a Tennessee contractor's license before? ______
   License Number ______________ Expiration Date ______________

5. If currently licensed in another state(s), indicate license number, limit and classifications:
   ____
   ____

6. List references with their complete mailing addresses:
   Licensed Architect or Engineer______________________
   ____
   Bank and Trust Company______________________
   ____
   Material Supply Dealer______________________
   ____
   Licensed Contractor______________________
   ____
   Past Client or Employer______________________
   ____

3

Figure 14-3: Typical application for electrical contractor's license

CONTRACTOR'S STATEMENT OF EXPERIENCE

Company Name ______________________ ☐ Corporation

Mailing Address ______________________ ☐ Partnership

☐ Individual

1. How many years has your company been in business as a contractor under your present business name? ____________
2. How many years experience has your company had:

   (a) As a Prime Contractor____________

   (b) As a Sub-Contractor____________

3. List a few representative projects your organization has completed during the past three years:

YEAR	TYPE OF WORK	CONTRACT AMOUNT	LOCATION OF WORK AND FOR WHOM PERFORMED

4. What types of work do you plan to perform as a Licensed Contractor?

5. Have you ever failed to complete any work awarded to you?____________If so, where and why?____________

6. Has any officer or partner of your organization ever been an officer or partner of some other oganization that failed to complete a construction contract?____________ If so, state name of individual, other organization, and reason therefor:

7. Has any officer or partner of your organization ever failed to complete a construction contract handled in his own name?____________ If so, state name of individual, name of owner, and reason therefor:

8. In what other lines of business are you financially interested?____________

9. If any of your assets or liabilities are related to any Parent, Subsidiary or Affiliated Company, list such Company, explain your connection with it, state where ownership lies and percentage of ownership:

NAME AND ADDRESS	EXPLANATION

4

Figure 14-3: Typical application for electrical contractor's license, continued

10. For what corporation or individuals have you performed work, and to whom do you refer?

11. For what cities or counties have you performed work, and to whom do you refer?

12. For what State Bureaus or Departments have you performed work, and to whom do you refer?

13. Have you ever performed work for the U. S. Government?__________ If so, when and to whom do you refer?

14. If an Individual or a Partnership, name the persons with whom you have been associated in business as partners or business associates in each of the last five years:

15. If you are requesting the classification of (s) asbestos material handling/removing, you must furnish evidence satisfactory to the Board that you have on staff at least one (1) designated employee who has completed an Asbestos Abatement Contractor's Training Course approved or sponsored by the United States Environmental Protection Agency (EPA). Provide the name of that employee and his/her position in your organization.

16. What is the construction experience of the principal individuals of your present organization? __________ years total

Individual's Name	Present Position or Office in Your Organization	Years of Construction Experience	Magnitude and Type of Work	In What Capacity

5

Figure 14-3: Typical application for electrical contractor's license, continued

EQUIPMENT QUESTIONNAIRE

1. Give names and addresses of all banks with whom you are currently doing business:

Name of Bank	Business Address	Bank Officer

2. Give names and addresses of material and equipment houses with whom you have established a line of credit:

Business Name	Business Address

3. Give names and addresses of firms or companies from whom you have purchased equipment during the last three years:

Business Name	Business Address

4. List the equipment you own and which is available to you by renting:

OWN	RENT

6

Figure 14-3: Typical application for electrical contractor's license, continued

Provide any additional information or remarks relating to questions 2, 3 and 4:

____
____
____
____
____
____
____

5. **(To be completed by highway contractors only.)** Are you familiar with the Standard Specifications for Road and Bridge construction adopted by the State of Tennessee, Department of Highways and Public Works, and approved by the U.S. Bureau of Public Roads which form the basis of proposals which are submitted?

____
____
____
____

Remarks: _______________________________________
____
____
____

List a few representative projects you have completed as a highway contractor and indicate the dollar amount of the project.

____
____
____
____
____
____
____
____
____
____
____
____
____
____
____

7

Figure 14-3: Typical application for electrical contractor's license, continued

- Most states will allow the examinees to use noiseless, nonrecording, battery-operated calculators and/or slide rules during the examination.
- All scratch work is usually permitted only on the blank spaces included in the examination booklet, NOT in the margins of the answer sheet(s). Answer sheets should contain only the required identifying information and responses to the examination questions.
- Permission of an examination proctor must be obtained before leaving the room while examination is in progress.
- Any examinee engaging in any kind of irregular conduct (such as giving or receiving help, conveying to others information about any questions appearing on the examination, using any materials not permitted, taking part in an act of impersonation, or removing examination materials or notes from the examination room) will be subject to disqualification and will be reported to the State Board of Examiners.

Description of the Examination

Master and Journeyman electrician exams, along with electrical contractor's exams can be open-book, multiple-choice, or a combination of open-book and closed-book multiple-choice. Some testing areas have no open-book exams. You are tested solely on what you know without referring to any reference material. Some questions may refer to a drawing, table or chart located either in the test booklet or in the *NEC* book.

Testing agencies also try to divide their questions into certain categories to ensure that most of the electrical field will be covered. The following lists the categories that appear on one state's Master Electrician's Exam. Most other areas will have similar coverage.

- ***Grounding and Bonding*** — This category makes up about 11% of the total test questions and covers determination of system and circuit grounding requirements, methods, and location of grounding connections. Choosing proper size grounding conductors, bonding of enclosures, equipment and metal piping systems.
- ***Services, Feeders, Branch Circuits, and Overcurrent Protection*** — These categories make up 13% of the total questions. Knowledge of the *NEC* rules covering services, electrical loads and determination of proper size and type of service and feeder conductors and ratings. Installation of panelboards, switchboards and overcurrent devices. Knowledge of circuit classifications, ratings, design and use requirements. Application of *NEC* rules covering electrical outlets, and devices, including wire connectors and methods.
- ***Raceways and Enclosures*** — These two categories make up 11% of the total exam questions and cover all types of raceways and their uses. Determining proper size conductor fill, support, and methods of installation. Application of proper type, use and support of boxes and cabinets.
- ***Conductors*** — About 9% of the exam questions cover electrical conductors. You will be asked to determine ampacity, type of insulation, usage, requirements, methods of installation, protection, support, and termination.
- ***Motors and Controls*** — 11% of the exam questions test your knowledge of *NEC* rules governing the installation of motors and motor controls. Includes calculations for motor feeders and branch circuits, short circuit, ground fault, and overload protection, and disconnecting means. Knowledge of all control circuits and motor type, application and usage.
- ***Utilization and General Use Equipment*** — You can expect about 11% of the total exam questions to cover this category. You should know the *NEC* rules covering

lighting, appliances, heating and air conditioning equipment, generators, transformers and the like.

- ***Special Occupancies/Equipment*** — About 6% of the exam questions will cover *NEC* rules that apply to hazardous locations, health care facilities, places of assembly, etc. You will also find questions on signs, welders, industrial machinery, and swimming pools.
- ***General Knowledge of the Electrical Trade and Calculations*** — As many as 25% of the exam questions may cover these topics. Terminology, practical calculations such as load computations, voltage drop, conductor derating, power factor, voltage and current ratings of equipment, and branch-circuit calculations.
- ***Low-Voltage Circuits Including Alarms and Communications*** — At least 3% of the total exam questions will test your knowledge of circuits and equipment operating at less than 50 volts, including all signal, alarm, and sound systems.

Please be aware that the above categories and percentages represent some testing agencies, but can vary from state to state.

Taking the Examination

In most states, the examination consists entirely of multiple-choice type questions. Read and understand all of the instructions carefully before attempting to answer any question. Reading the instructions too fast or skipping over any part may cause you to miss something important and possibly arrive at an incorrect answer.

Keep track of time: Do not spend too much time on any one question. If a question is difficult for you, mark on the answer sheet the answer you think is correct and place a check by that question in the examination booklet. Then go on to the next question; if you have time after finishing the rest of the examination you can go back to the questions you have checked.

Your answers to the questions on the examination are usually recorded on respective separate answer sheets provided. See Figure 14-4. Answer spaces on the answer sheets are lettered to correspond with the letters of the possible answers printed in the examination booklet. For each question, you are to decide which one of the four possible answers is best and blacken the appropriate lettered space on your answer sheet. The following example illustrates how all answers usually are marked on your qualifying examination answer sheet.

Which of the following is the basic unit used to measure current flow?

(A) Amperes

(B) Volts

(C) Ohms

(D) Watts

(Since choice A is the best answer, the A space is to be blackened.)

When marking answers, follow these instructions:

- Do not use ink or ballpoint pen
- Use black lead #2 pencil only
- Make heavy marks that fill the circle completely
- Erase cleanly any answer you wish to change
- Make no stray marks on the answer sheet

If you mark more than one answer to any question by darkening more than one lettered space, it will be graded as incorrect. Therefore, if you change an answer, be sure that any previous marks for that question are erased completely.

Your grade on the examination will be determined by the total number of questions you answer ***correctly***. Do the best you can. Since very few examinees answer ***all*** questions ***correctly***, do not be concerned if there are a few you cannot answer.

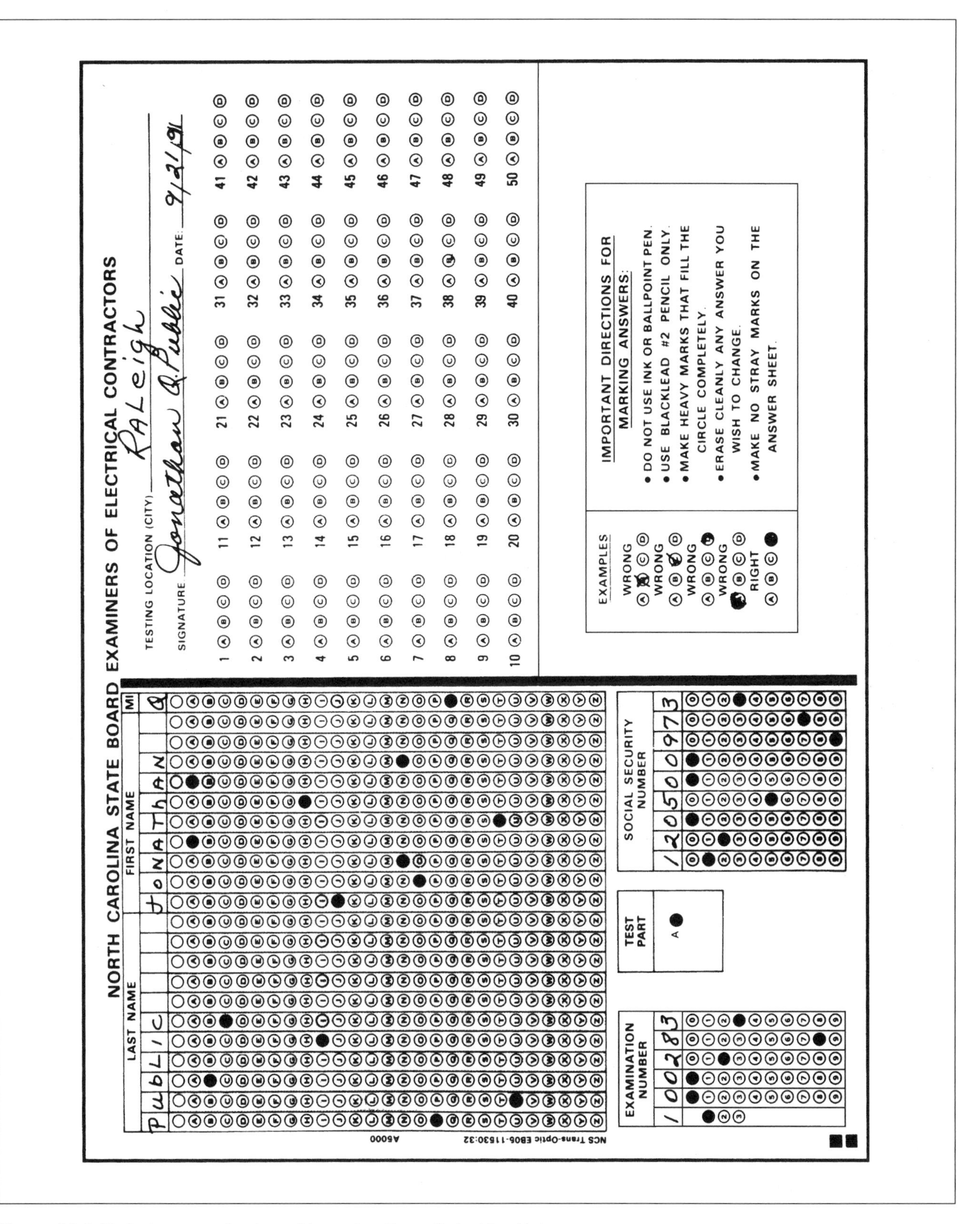

NORTH CAROLINA STATE BOARD EXAMINERS OF ELECTRICAL CONTRACTORS

TESTING LOCATION (CITY) RALeigh

SIGNATURE Jonathan Q. Public DATE 9/21/91

LAST NAME: PUbLIC

FIRST NAME: JONAThAN

MI: Q

EXAMINATION NUMBER: 100283

TEST PART: A

SOCIAL SECURITY NUMBER: 120500973

1–50 Ⓐ Ⓑ Ⓒ Ⓓ

EXAMPLES

WRONG

WRONG

WRONG

WRONG

RIGHT

IMPORTANT DIRECTIONS FOR MARKING ANSWERS:

- DO NOT USE INK OR BALLPOINT PEN.
- USE BLACKLEAD #2 PENCIL ONLY.
- MAKE HEAVY MARKS THAT FILL THE CIRCLE COMPLETELY.
- ERASE CLEANLY ANY ANSWER YOU WISH TO CHANGE.
- MAKE NO STRAY MARKS ON THE ANSWER SHEET.

NCS Trans-Optic EB06-11530:32 A5000

Figure 14-4: Typical answer sheet used in conjunction with text booklets

If you have some knowledge of a question, even though you are uncertain about the answer, you may be able to eliminate one or more of the answer choices that are wrong. In such cases, it is better to guess at the correct answer rather than leave the answer space blank.

Sample Questions

All of the questions appearing in this book are designed to illustrate the type of questions that appear on electrician's exams throughout the country. But let's review the exact process of answering questions that appear on these exams. Keep in mind that most questions will not have supplemental illustrations; other questions will have illustrations to refer to when answering the question. In this case, make sure you look at the correct illustrations as they appear in different locations in the exam booklet; that is, an illustration might appear above the question, or it might appear below it.

Example 1: Conductor P in Figure 14-5 must be identified by which of the following colors?

(A) White

(B) Gray

(C) Black

(D) Green

Here you are asked to select from the listed colors the one that is to be used to identify the equipment grounding conductor of a branch circuit. Since *NEC* Section 250.119 requires that green or green with yellow stripes be the color of insulation used on a grounding conductor (when it is not bare), the correct answer is (D).

Example 2: A circuit leading to a gasoline dispensing pump must have a disconnecting means:

(A) only in the ungrounded conductors

(B) only in the grounded (neutral) conductor

(C) operating independently in all conductors

(D) operating simultaneously in all conductors

Here the "question" is in the form of an incomplete statement. Your task is to select the choice that best completes the statement. In this case, you should have selected (D) since Section 514.11(A) of the *NEC* specifies that such a circuit shall be provided with a means to disconnect simultaneously from the source of supply all conductors of the circuit, including the grounded neutral, if any.

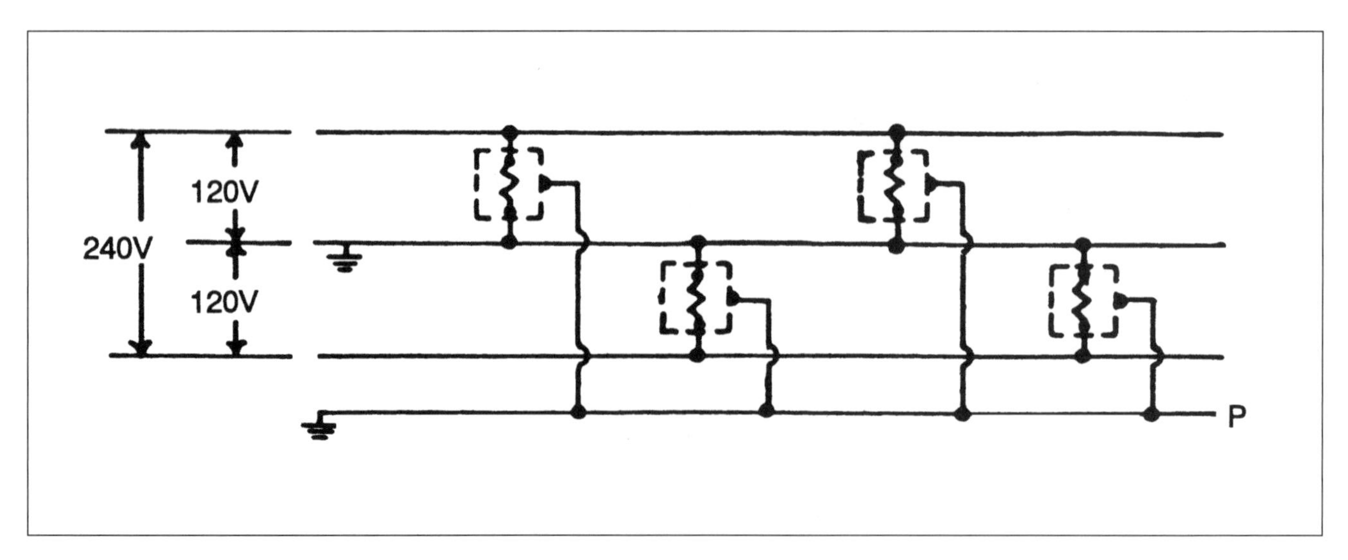

Figure 14-5: A 120/240-volt branch circuit with grounding conductor

Example 3: If the grounded (neutral) conductor of a dryer circuit is used to ground the frame of an electric clothes dryer, all of the following conditions must be met ***except*** when the:

(A) the supply circuit is a 120/240V, single-phase, 3-wire circuit derived from a three-phase, 4-wire delta system

(B) the grounded (neutral) conductor is not smaller than #10 copper or #8 aluminum

(C) the grounded (neutral) conductor is insulated, or if uninsulated, part of a Type SE service entrance cable and the branch circuit originates at the service equipment

(D) the grounding contacts of the receptacle installed with the dryer are bonded to the dryer

Again the "question" is in the form of an incomplete statement and your task is to select the choice that best completes the statement. In this case, you are to find an exception. You have to select the condition that does *not* have to be met when using the grounded (neutral) conductor of the circuit to ground the frame of the dryer. You should have selected (A) because *NEC* Section 250-140 requires the conditions listed in (B), (C), and (D) but does not require or permit the conditions listed in (A).

Example 4: Type TW copper conductors in a raceway are used for feeder conductors in the diagram shown in Figure 14-6. Which of the following is the minimum allowable size of these feeder conductors?

(A) No. 8

(B) No. 6

(C) No. 4

(D) No. 2

Here you are asked to determine the smallest

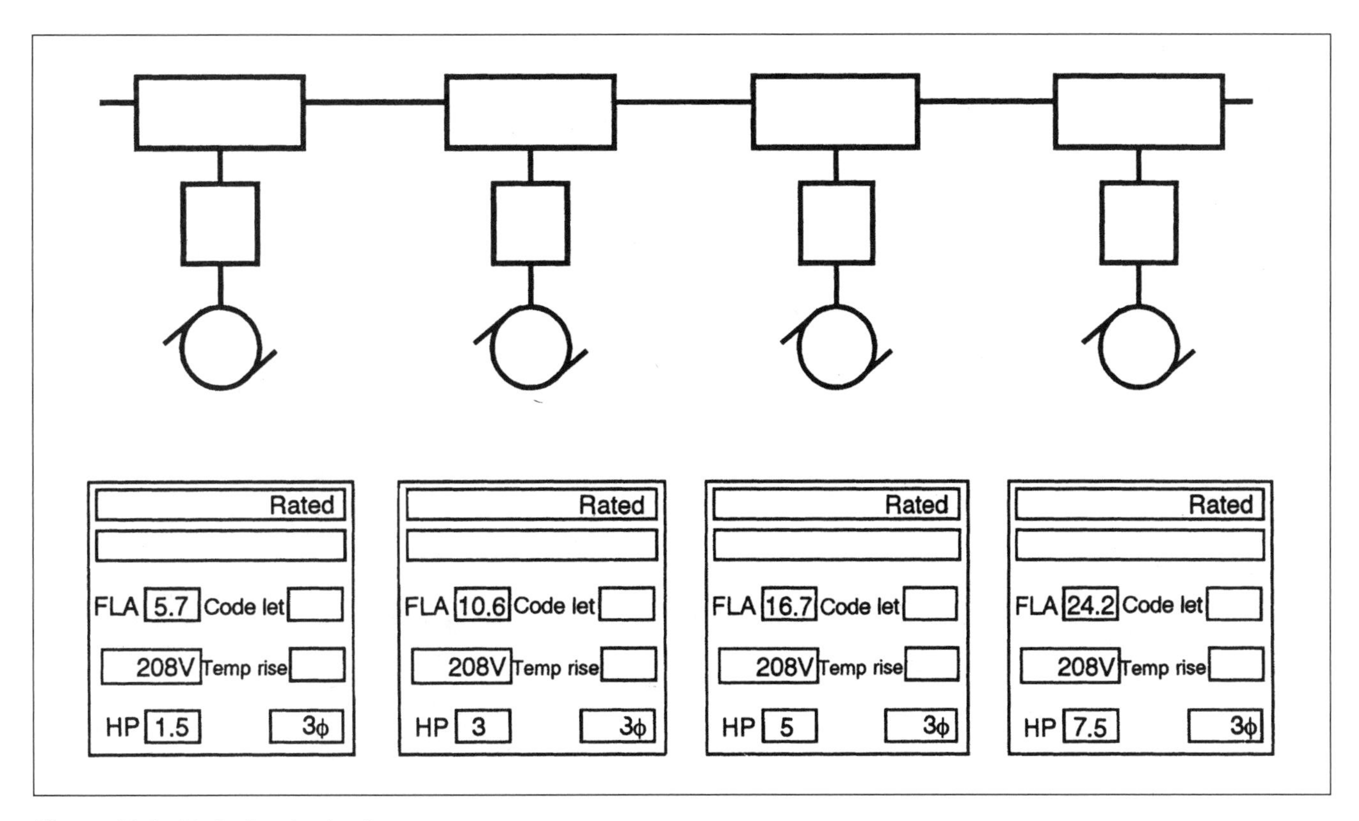

Figure 14-6: Typical motor feeder

size of conductor that the *NEC* allows for the motor feeder. *NEC* Section 430.24 tells you to add together the full-load current rating of the four motors plus 25% of the rating for the largest motor. In this case, the total is 63 amperes. This figure is then used to enter *NEC* Table 310.16; and according to the Table, No. 4 TW copper conductor has a rated ampacity of 70 amperes; therefore, you should have selected (C).

Example 5: Which of the following conditions shall be met to omit overcurrent protection at the tap point for a 20-ft. feeder tap? I. The ampacity of the tap conductors shall be at least one-half the ampacity of the feeder conductors. II. The tap conductor shall terminate in a panelboard.

(A) I only

(B) II only

(C) Both I and II

(D) Neither I nor II

Here you are asked to determine which of two conditions are required by the *NEC* if a contractor omits an overcurrent device at a tap point. Exception No. 3 to *NEC* Section 240.21(B)(2) does list four conditions that must be met when considering such an omission. However, the question asks only about the two conditions specifically stated and, since neither of them is included in the exception as a condition, you should have answered D.

Final Examination

The examination to follow is meant to test your understanding of electrical theory and application. The questions are representative of typical Master Electrician's and Electrical Contractor's examinations given around the United States for both city and state licenses. Complete solutions to all examination problems are contained in Appendix II.

Before taking this exam, however, make sure you have thoroughly studied the questions and answers in the previous 13 chapters. Then answer the questions in this exam as if you were actually taking an exam to obtain your Master Electrician's or State Contractor's license. This might be the only chance you have to take a sample exam prior to taking the real thing.

Instructions: Obtain several pieces of blank lined paper such as a legal pad. Write "Morning Exam" at the top of one page, and then number the lines 1 through 100. This sheet of paper will be for your answers to the questions. Or you can photocopy the computer answer form in this book (Figure 14-7) and follow the instructions on the form. Do not write your answers in this book. The morning exam should not take you longer than four hours to complete. To simulate an actual situation, you might want to wake up on your day off from work (like on Saturday morning), eat a good breakfast, have your numbered papers and several pencils handy, and start the exam as if you are at your state licensing location. Find a quiet location in your home and ask your family to cooperate. Then complete the exam without any references.

If you cannot spend a full four hours on the exam at one sitting, keep account of the time actually spent, and then return to the exam later. But do not spend more than a total of four hours on the morning exam. Once completed, put the answers to the morning exam in a safe place. Do not grade your exam yet.

Now take another sheet of paper and write "Afternoon Exam" at the top of the page and again number the lines 1 to 100 for the answers to the "Afternoon Exam." When time permits, answer the questions on the "Afternoon Exam" as you did for the "Morning Exam."

While taking either of these two examinations, if you don't know the answer to a question, don't stop taking the exam; remember you are working under a time limit. Go on to the next question. If you have time when you have gone through all the questions, then go back to ones you did not answer and try to think of the solution. However, don't look up the answer. Sure, you can cheat now and score high on this exam, but you won't have that opportunity when the actual examination day comes.

When both exams are completed, have someone else grade your answer sheets, if possible. Remember, 70% is the lowest passing grade in most states. This means you must correctly answer 140 out of a total of 200 questions to pass.

The person grading your exam should also write the reference notes to those questions missed. This will give you a quick-reference for further study in the fields you find that you are the weakest in.

Now, let's assume that you made, say, 72% on this sample exam, which is a passing grade. Are you ready for the real thing? You might be, but I wouldn't chance it. You will not have these same questions on the real exam, and there might be more questions on the real exam in your weakest area. Therefore, it is recommended that you do further study in the field or fields in which you are the weakest. Wait a few days and then take the exam again — following the same procedure as before. Make sure you keep the answers from your previous attempt out of sight.

If you score 90% or more on this sample test, you stand a good chance of passing the real electrician's or electrical contractor's exam. But do you want to stop here? Remember, virtuoso musicians always practice techniques more difficult or beyond their performance range. By doing so, the ranges in which they actually perform in public come easier. By the same token, the closer you can score to 100% on this sample exam, the better your chances of passing the real examination, and passing the real examination is the reason you have this book. So be serious and good luck!

Electrician's Exam Answer Sheet

Name ______________________________

Please print (last) (first) (middle)

Address ______________________________

____

Signature ______________________________

1 (A) (B) (C) (D)	26 (A) (B) (C) (D)	51 (A) (B) (C) (D)	76 (A) (B) (C) (D)
2 (A) (B) (C) (D)	27 (A) (B) (C) (D)	52 (A) (B) (C) (D)	77 (A) (B) (C) (D)
3 (A) (B) (C) (D)	28 (A) (B) (C) (D)	53 (A) (B) (C) (D)	78 (A) (B) (C) (D)
4 (A) (B) (C) (D)	29 (A) (B) (C) (D)	54 (A) (B) (C) (D)	79 (A) (B) (C) (D)
5 (A) (B) (C) (D)	30 (A) (B) (C) (D)	55 (A) (B) (C) (D)	80 (A) (B) (C) (D)
6 (A) (B) (C) (D)	31 (A) (B) (C) (D)	56 (A) (B) (C) (D)	81 (A) (B) (C) (D)
7 (A) (B) (C) (D)	32 (A) (B) (C) (D)	57 (A) (B) (C) (D)	82 (A) (B) (C) (D)
8 (A) (B) (C) (D)	33 (A) (B) (C) (D)	58 (A) (B) (C) (D)	83 (A) (B) (C) (D)
9 (A) (B) (C) (D)	34 (A) (B) (C) (D)	59 (A) (B) (C) (D)	84 (A) (B) (C) (D)
10 (A) (B) (C) (D)	35 (A) (B) (C) (D)	60 (A) (B) (C) (D)	85 (A) (B) (C) (D)
11 (A) (B) (C) (D)	36 (A) (B) (C) (D)	61 (A) (B) (C) (D)	86 (A) (B) (C) (D)
12 (A) (B) (C) (D)	37 (A) (B) (C) (D)	62 (A) (B) (C) (D)	87 (A) (B) (C) (D)
13 (A) (B) (C) (D)	38 (A) (B) (C) (D)	63 (A) (B) (C) (D)	88 (A) (B) (C) (D)
14 (A) (B) (C) (D)	39 (A) (B) (C) (D)	64 (A) (B) (C) (D)	89 (A) (B) (C) (D)
15 (A) (B) (C) (D)	40 (A) (B) (C) (D)	65 (A) (B) (C) (D)	90 (A) (B) (C) (D)
16 (A) (B) (C) (D)	41 (A) (B) (C) (D)	66 (A) (B) (C) (D)	91 (A) (B) (C) (D)
17 (A) (B) (C) (D)	42 (A) (B) (C) (D)	67 (A) (B) (C) (D)	92 (A) (B) (C) (D)
18 (A) (B) (C) (D)	43 (A) (B) (C) (D)	68 (A) (B) (C) (D)	93 (A) (B) (C) (D)
19 (A) (B) (C) (D)	44 (A) (B) (C) (D)	69 (A) (B) (C) (D)	94 (A) (B) (C) (D)
20 (A) (B) (C) (D)	45 (A) (B) (C) (D)	70 (A) (B) (C) (D)	95 (A) (B) (C) (D)
21 (A) (B) (C) (D)	46 (A) (B) (C) (D)	71 (A) (B) (C) (D)	96 (A) (B) (C) (D)
22 (A) (B) (C) (D)	47 (A) (B) (C) (D)	72 (A) (B) (C) (D)	97 (A) (B) (C) (D)
23 (A) (B) (C) (D)	48 (A) (B) (C) (D)	73 (A) (B) (C) (D)	98 (A) (B) (C) (D)
24 (A) (B) (C) (D)	49 (A) (B) (C) (D)	74 (A) (B) (C) (D)	99 (A) (B) (C) (D)
25 (A) (B) (C) (D)	50 (A) (B) (C) (D)	75 (A) (B) (C) (D)	100 (A) (B) (C) (D)

Figure 14-7: Answer sheet for use with morning exam

MORNING EXAM
TIME: 4 HOURS

1. A lighting fixture intended for installation in a metal forming shell mounted in a pool or fountain structure where the fixture will be completely surrounded by water is called:

A) Dry-niche fixture
B) Plug-connected lighting assembly
C) Wet-niche fixture
D) Cord-connected lighting assembly

2. What is the minimum distance that receptacles can be installed from the inside walls of a swimming pool?

A) 8 feet
B) 3 feet
C) 6 feet
D) 10 feet

3. In which of the following locations is induction heating equipment prohibited?

A) Hazardous (classified) locations
B) Industrial structures
C) Commercial buildings
D) Multifamily dwellings

4. An assembly of electrically interconnected electrolytic cells supplied by a source of direct-current power is known as:

A) Cellular circuit
B) Intercell
C) Cell line
D) Electrolytic cellular communication

5. The power used by a load supplied by a 20-ampere, 120-volt branch circuit cannot exceed:

A) 2400 volt-amperes
B) 2000 volt-amperes
C) 5000 volt-amperes
D) 4600 volt-amperes

6. The maximum continuous load connected to a 20-ampere branch must not exceed:

A) 20 amperes
B) 16 amperes
C) 14 amperes
D) 12 amperes

7. The unit lighting load for a store building is:

A) 3 volt-amperes/square foot
B) 5 volt-amperes/square foot
C) 2 volt-amperes/square foot
D) 6 volt-amperes/square foot

8. The ampacity of conductors connecting a power-factor correcting capacitor to a motor circuit shall not be less than:

A) 135% of the rated current of the capacitor

B) 110% of the feeder capacity

C) 135% of the ampacity of the motor circuit conductors

D) One-third the ampacity of the motor circuit conductors and not less than 135% of the rated current of the capacitor

9. If the phase-to-neutral voltage in a three-phase system is 2400 volts, the phase-to-phase voltage is approximately:

A) 4800

B) 4150

C) 2400

D) 12,470

10. The rating of the electric service for a one-family residence shall not be less than:

A) 100 amps

B) 100 amps when the initial load is 10 kW or less

C) 150 amps when the initial load is 15 kW or less

D) 60 amps

11. When it is allowed, the demand factor applied to the service neutral for a load in excess of 200 amperes is:

A) 50%

B) 60%

C) 70%

D) 90%

12. The power used by a load supplied by a 15-ampere, 120-volt branch circuit cannot exceed:

A) 2400 volt-amperes

B) 2000 volt-amperes

C) 1800 volt-amperes

D) 4600 volt-amperes or 4300 watts on residential circuits

13. The minimum size conductors required for a 15-ampere branch circuit using Type NM cable are:

A) No. 10 AWG aluminum

B) No. 12 AWG copper

C) No. 14 AWG copper

D) No. 8 AWG aluminum or copper-clad aluminum conductors

14. The maximum continuous load connected to a 15-ampere branch must not exceed:

A) 15 amperes
C) 20 amperes
B) 12 amperes
D) 16 amperes

15. If a single-phase motor draws 50 amperes and its branch circuit is protected by a 150-ampere nontime-delay fuse, the feeder overcurrent protective device may not be larger than:

A) 87.5 amperes
C) 300 amperes
B) 100 amperes
D) 150 amperes

16. If the standard calculation method is used, the feeder demand for a dwelling with a 150-kilovolt-ampere general lighting load is:

A) 52,500 volt-amperes
C) 54,450 volt-amperes
B) 56,450 volt-amperes
D) 51,450 volt-amperes

17. If the optional service calculation method for a dwelling unit is used, the load of less than four separately controlled electric space-heating units is subject to a demand factor of:

A) 100%
C) 80%
B) 65%
D) 40%

18. If the optional calculation method for a one-family dwelling is used, all "other loads" above the initial 10 kilowatts are subject to a demand factor of:

A) 50%
C) 40%
B) 80%
D) 65%

19. A branch circuit supplies a household electric range with a demand of 8000 watts. The ungrounded conductors have an ampacity of 40 amperes. The neutral ampacity must be at least:

A) 40 amperes
C) 35 amperes
B) 28 amperes
D) 24 amperes

20. What is the minimum size fixture wire permitted for the control and operating circuits of X-ray and auxiliary equipment?

A) 22

C) 18

B) 14

D) 12

21. There is no overcurrent protection on the secondary side of a 480-volt transformer with a primary current of 100 amperes. The rating of the primary overcurrent protective device should not exceed:

A) 100 amperes

C) 125 amperes

B) 250 amperes

D) 60 amperes

22. Receptacle outlets in a dwelling must be installed in habitable rooms so that no point along the floor line is farther from an outlet than:

A) 12 feet

C) 4 1/2 feet

B) 6 feet

D) 5 feet

23. The wall space between two doors in a living room requires a receptacle outlet if the space is wider than:

A) 3 feet

C) 4 1/2 feet

B) 4 feet

D) 2 feet

24. Ground-fault circuit-interrupter protection is required for all 120-volt, 15- or 20-ampere receptacles installed in the following dwelling unit area:

A) Bedroom

C) Bath

B) Living room

D) Hallway

25. A receptacle outlet is required above a show window every:

A) 3 feet

C) 1 1/2 feet

B) 12 feet

D) 6 feet

26. A 120-volt circuit has an electric heater connected with a current rating of 7.5 amperes. What is the resistance in ohms of the connected pure resistance load?

A) 8 ohms
B) 10 ohms
C) 16 ohms
D) 20 ohms

27. An incandescent lamp has a resistance of 104 ohms when 2 amperes of current flows. What is the voltage?

A) 240 volts
B) 208 volts
C) 120 volts
D) 12 volts

28. What is the minimum general lighting load permitted by the *NEC* in a 4500 square foot warehouse?

A) 1005 volt-amperes
B) 1125 volt-amperes
C) 1750 volt-amperes
D) 1243 volt-amperes

29. The voltage per turn for a transformer is 1.25 volts. What is the voltage of the transformer if it has 192 turns?

A) 12 volts
B) 240 volts
C) 120 volts
D) 24 volts

30. A 480/277-volt, Y-connected transformer is used to supply a balanced 277-volt, single-phase lighting load of 40,000 watts. What size transformer (kVA) should be used?

A) 10 kVA
B) 20 kVA
C) 40 kVA
D) 50 kVA

31. What is the maximum allowable wattage that can be connected to a 240-volt, 20-ampere, single-phase circuit feeding residential electric heaters?

A) 3840 watts
B) 4000 watts
C) 4500 watts
D) 1920 watts

32. What is the approximate wattage of a 240-volt motor with a resistance of 28 ohms?

A) 1780 watts

B) 2057 watts

C) 3640 watts

D) 1045 watts

33. Heavy-duty type lampholders are required on branch circuits having a rating in excess of:

A) 20 amperes

B) 30 amperes

C) 40 amperes

D) 15 amperes

34. Receptacle outlets installed in bathrooms of dwelling units always require:

A) Twist-lock receptacles

B) Ground-fault circuit-interrupter protection

C) Nongrounded receptacles

D) Waterproof receptacle covers

35. Insulated conductors size No. 6 or smaller intended for use as a grounded conductor in a circuit must be which of the following color(s)?

A) Continuous white

B) Gray

C) Three white stripes along the entire length of other than green insulation

D) All of the above

36. Cable tray systems must *not* be used:

A) For power and control applications

B) For service-entrance systems

C) For signal cables

D) In hoistways

37. EMT shall be supported at least every:

A) 4 feet

B) 8 feet

C) 10 feet

D) 15 feet

38. All components between the point of termination of the overhead service drop or underground service lateral and the building main disconnecting device, with the exception of the power company's metering equipment, are called:

A) Service entrance

B) Service-entrance conductors

C) Service drop

D) Service-entrance equipment

39. On a 240-volt, single-phase, 3-wire system, the neutral conductor will:

A) Never carry current even when the other conductors carry a different amount

B) Never carry current larger than the difference between the current in the two "hot" legs

C) Carry current equal to the current between phases

D) Carry current equal to 1/2 the current between phases

40. The minimum allowable current rating of 3-wire service-entrance conductors for a residence with six or more 2-wire branch circuits is:

A) 60 amperes

B) 100 amperes

C) 125 amperes

D) 150 amperes

41. What is the smallest aluminum or copper-clad wire size allowed for service-entrance conductors supplying loads consisting of not more than two 2-wire branch circuits?

A) No. 10 AWG

B) No. 8 AWG

C) No. 6 AWG

D) No. 4 AWG

42. To measure the area of a dwelling to determine the lighting load, the following dimensions are used:

A) The floor area computed from the inside dimensions

B) The cubic feet of each room

C) The floor area computed from the outside dimensions

D) The area 6 inches from the inside walls

43. A 400-ampere electric service is normally metered with a combination of a watt-hour meter and:

A) Current transformers

B) VAR meters

C) Capacitors

D) Ammeter

44. Locations where fibers are stored are:

A) Class I, Division 1

B) Class II, Division 1

C) Class III, Division 1

D) Class III, Division 2

45. The disconnecting means for a 120/240-volt circuit for a gasoline dispenser must disconnect:

A) The neutral

B) All conductors

C) The grounded conductor

D) The ungrounded conductors

46. The branch-circuit overcurrent device may serve as the disconnecting means for motors rated less than:

A) 20 horsepower

B) 1/8 horsepower

C) 15 horsepower

D) 5 horsepower

47. A disconnecting means for a 2300-volt motor must be capable of being locked in the:

A) Closed position

B) Open and closed positions

C) Open position

D) "ON" position

48. A site plan is a plan view (as if viewed from an airplane) that shows:

A) Each floor level of the building

B) Power-riser diagrams

C) Cross-sections of the building

D) Property boundaries and buildings

49. A legend or electrical symbol list is shown on electrical working drawings to:

A) Describe materials and installation methods

B) Show the outline of the architect's floor plan

C) Identify all symbols used to indicate electrical outlets or equipment

D) Enable the electric service size to be calculated

50. All outside receptacles installed in a residence or dwelling must have the following:

A) Double-pole circuit breaker

B) A ground-fault protector (ground-fault circuit-interrupter)

C) A single-pole mercury switch

D) A 40-amp circuit breaker

51. All outside receptacles installed in a residental garage must have the following:

A) Double-pole circuit breaker

B) A ground-fault circuit-interrupter

C) A single-pole mercury switch

D) A 40-amp circuit breaker

52. The service grounding conductor is sized by the rating of:

A) The overcurrent protective device

B) The service-entrance conductors

C) The supply transformer

D) The load to be served

53. A 120/240-volt, 200-ampere service requires a neutral with a minimum ampacity of:

A) 140 amperes

B) 160 amperes

C) 175 amperes

D) 200 amperes

54. In a multiple-occupancy building, each occupant shall:

A) Not have access to the occupant's service disconnecting means

B) Have access to the occupant's service disconnecting means

C) Have access to all occupant's service disconnecting means

D) Have complete access to all electrical equipment

55. Although enclosed, a service disconnecting means must still be:

A) Locked securely

B) Always kept unlocked

C) Locked only while operable

D) Externally operable

56. Name one approved method to disconnect the grounded conductor from the premises' wiring.

A) A terminal with pressure connectors

B) A ground clamp on a grounding electrode

C) A wire trough

D) A cablebus

57. The minimum allowable current rating of 3-wire service-entrance conductors for a residence with an initial net computed load of 6 kVA or more is:

A) 100 amperes

B) 60 amperes

C) 125 amperes

D) 150 amperes

58. What is the purpose of drip loops in service conductors?

A) To insure a tight connection

B) To impede the progress of termites

C) To help keep water out of the service head

D) For easier connection to the service-entrance conductors

59. What is the dimension of "A" on the architect's scale (1 inch = 1 foot) in Figure 14-8?

A) 6 inches
B) 1 foot
C) 5 inches
D) 5 feet

60. What is the dimension of "B" on the architect's scale in Figure 14-8?

A) 1 foot
B) 17 feet
C) 2 feet
D) 3 feet

61. What is dimension "C" on the architect's scale in Figure 14-8?

A) 6 feet
B) 6 feet 6 inches
C) 9 feet
D) 4 feet

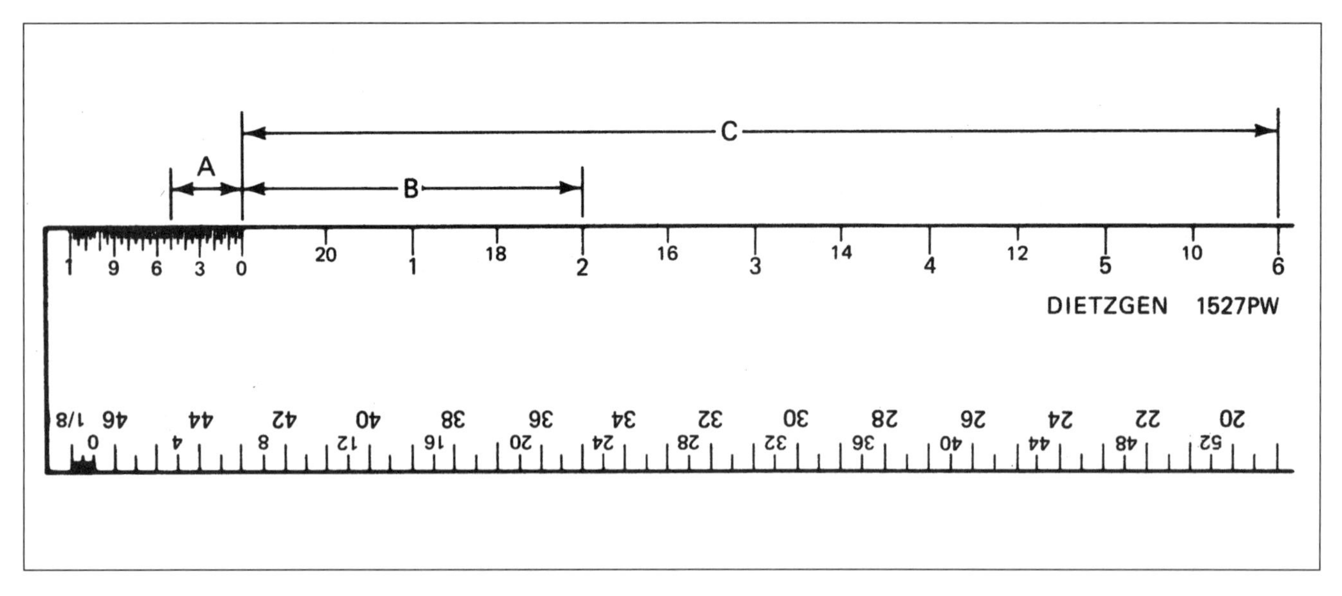

Figure 14-8: Architect's scale

62. Switches or similar devices controlling a transformer in a sign must have a rating based on what percent of the ampere rating of the transformer?

A) 100%
C) 125%
B) 200%
D) 300%

63. A branch circuit supplying lamps in a sign must not be rated more than:

A) 20 amperes
C) 30 amperes
B) 15 amperes
D) 40 amperes

64. Wood used for decoration on a sign must not be closer to a lampholder than:

A) 6 inches
C) 3 inches
B) 1 foot
D) 2 inches

65. What is the smallest size conductor permitted for wiring neon secondary circuits rated at 1000 volts or less?

A) No. 14 AWG
C) No. 16 AWG
B) No. 12 AWG
D) No. 18 AWG

66. The smallest size branch circuit required for wiring a neon tubing installation is:

A) 20 amp
C) 30 amp
B) 15 amp
D) 50 amp

67. Branch circuits that supply signs shall not exceed:

A) 15 amps
C) 30 amps
B) 20 amps
D) 50 amps

68. Cords supplying portable gas tube signs must have a maximum length of:

A) 6 feet
C) 12 feet
B) 15 feet
D) 3 feet

69. Which of the following wiring methods is not suitable for underfloor wiring in a computer room?

A) Type MI cable
B) Rigid conduit
C) EMT
D) Type NM cable

70. What is the minimum depth of clear working space at electrical equipment rated from 601 to 2500 volts with exposed live parts on one side and no live or grounded parts on the other?

A) 5 feet
B) 8 feet
C) 3 feet
D) 6 feet

71. The term "accessible" as applied to wiring methods means:

A) Admitting close approach
B) Capable of being removed or exposed without damaging the building structure or finish
C) Capable of being reached quickly for operation, renewal, or inspection
D) Within sight of the operator

72. The term "concealed" as applied to wiring methods means:

A) Admitting close approach
B) Capable of being removed or exposed without damaging the building structure or finish
C) Capable of being reached quickly for operation, renewal, or inspection
D) Rendered inaccessible by the structure of the building

73. A bathroom is defined by the *NEC* as a room or area with a wash basin and:

A) A toilet
B) A ground-fault circuit-interrupter
C) A locking door
D) An exhaust fan

74. Branch-circuit conductors supplying one or more pieces of electroplating equipment must have a minimum ampacity of:

A) 125% of the total connected load
B) 150% of the total connected load
C) 80% of the total connected load
D) 200% of the total connected load

75. A feeder is defined as:

A) A circuit conductor between the final overcurrent device protecting the circuit and the outlet

B) A branch circuit that supplies several outlets

C) An apparatus for generating electricity

D) Circuit conductors between the service, or the source of a separately derived system, and the final branch-circuit overcurrent device

76. A branch circuit is defined as:

A) Circuit conductors between the service, or the source of a separately derived system, and the final branch-circuit overcurrent device

B) A branch circuit that supplies several outlets

C) An apparatus for generating electricity

D) A circuit conductor between the final overcurrent device protecting the circuit and the outlet

77. Which of the following statements about supplementary overcurrent protection is correct?

A) It must not be used in lighting fixtures

B) It may be used as a substitute for a branch-circuit overcurrent device

C) It may be used to protect internal circuits of equipment

D) It must be readily accessible

78. Which of the following statements about Type MI cable is correct?

A) It may be used in any hazardous (classified) location

B) It must be supported and secured at intervals not exceeding 10 feet

C) A single run of cable must not contain more than the equivalent of 3 quarter bends

D) It must not be used where exposed to oil and gasoline

79. The maximum allowable ampacity of each of three (3) single insulated 75°C rated No. 6 AWG copper conductors in a raceway is:

A) 55 amperes

B) 70 amperes

C) 65 amperes

D) 80 amperes if run in PVC conduit under ground at a depth of 12 inches or more

80. Which of the following statements about the protection of nonmetallic-sheathed cable from physical damage is/are correct? I. When passing through a floor, the cable shall be enclosed in metal pipe or conduit extending at least 6 inches above the floor. II. When run across the top of the floor joists in an accessible attic, the cable shall be protected by guard strips.

A) Neither I nor II
B) Both I and II
C) I only
D) II only

81. Which of the following statements about a single-throw knife switch is/are correct? I. The switch shall be placed so that gravity will not tend to close the blades. II. A switch approved for inverted mounting shall be provided with a locking device that will ensure that the blades remain in the open position when so set.

A) I only
B) II only
C) Both I and II
D) Neither I nor II

82. Which of the following statements about a No. 2 THHN copper conductor is correct?

A) Its maximum operating temperature is 90°C
B) It has nylon insulation
C) Its area is 0.067 inches
D) It has a dc resistance of .26 ohms per foot

83. A lighting and appliance branch circuit panelboard contains six (6) 3-pole circuit breakers and eight (8) 2-pole circuit breakers. The maximum allowable number of single-pole circuit breakers permitted to be added to this panelboard is:

A) 8
B) 16
C) 6
D) 42

84. A metal device box contains cable clamps, six (6) No. 14 AWG conductors, and one (1) single-pole dimmer switch. Which of the following is the minimum allowable box size?

A) 12 cubic inches
B) 13.5 cubic inches
C) 14.25 cubic inches
D) 18 cubic inches

85. A 120/240-volt, 3-wire service drop passing over a residential driveway shall have a minimum height above grade of:

A) 10 feet
B) 12 feet
C) 15 feet
D) 18 feet

86. The minimum size liquidtight flexible metal conduit is:

A) 3/4 inch

B) 3/8 inch

C) 1 inch

D) 1 1/2 inch

87. A metal switch box with cable clamps contains two (2) No. 14/2 nonmetallic-sheathed cables with ground. Which of the following is the minimum allowable capacity of the box if no other devices are present?

A) 14 cubic inches

B) 16 cubic inches

C) 12 cubic inches

D) 18 cubic inches

88. The rating of the overcurrent device supplying a branch circuit specifically installed for the purpose of supplying a low-voltage wiring system is rated at fifteen (15) amperes. The minimum size equipment grounding conductor required for this dedicated circuit is:

A) No. 14 aluminum

B) No. 12 copper

C) No. 14 copper

D) No. 10 aluminum

89. The grounded (neutral) conductor used in a branch circuit must be identified by an outer covering of any of the following except:

A) Gray

B) White with a green stripe

C) White

D) Three continuous white stripes on other than green insulation

90. An individual branch circuit is rated at 20 amperes and serves a single receptacle in a single-family dwelling. What is the required minimum rating of the receptacle if this is the only load on the circuit?

A) 15 amperes

B) 20 amperes

C) 25 amperes

D) 30 amperes

91. A 240-volt, single-phase branch circuit supplying a 21.85 kW commercial cooking appliance has copper conductors with Type TW insulation. Which of the following is the minimum allowable conductor size?

A) No. 3 AWG

B) No. 2 AWG

C) No. 1 AWG

D) No. 1/0

92. A store building measures 60 feet by 80 feet and is supplied by a 120/240-volt, three-phase, 4-wire delta system; the building has 60 linear feet of show-window lighting. Which of the following is the minimum allowable number of 2-wire, 20-ampere branch circuits required to supply the continuous general lighting and show-window lighting in this building?

A) 7
B) 8
C) 11
D) 14

93. A 75 kVA transformer has a 480-volt, three-phase delta primary and a 120/208-volt, three-phase, 4-wire wye secondary; it is fed from a 225-ampere plug-in bus protected by a 225-ampere circuit breaker. Which of the following statements about overcurrent protection for this transformer is/are correct? I. It shall have a primary overcurrent device rated or set not more than 125 amperes, regardless of the rating of its secondary overcurrent device. II. If no additional overcurrent device is provided on the primary side, it shall have an overcurrent device on the secondary side rated or set at not more than 300 amperes.

A) I only
B) II only
C) Both I and II
D) Neither I nor II

94. A branch circuit supplies a single hermetic refrigerant motor-compressor for an air conditioning unit in a single-family dwelling. The hermetic refrigerant motor-compressor rated-load current is 18 amperes. If a 30-ampere fuse will not carry the motor-compressor, the maximum rating of the branch-circuit overcurrent-protective device may be increased to a maximum of:

A) 35 amperes
B) 40 amperes
C) 45 amperes
D) 50 amperes

95. The minimum allowable number of 120-volt, 15-ampere, 2-wire lighting branch circuits required for a residence 70 feet by 30 feet are:

A) 2
B) 3
C) 4
D) 5

96. Which of the following may be used as a feeder from the service equipment to a mobile home? I. A permanently installed circuit calculated to comply with applicable *NEC* requirements. II. Not more than one 50-ampere power-supply cord.

A) I only
B) II only
C) Neither I nor II
D) Either I or II

97. Which of the following statements about 120/240-volt 3-wire service-entrance conductors for a one-family dwelling is/are correct? I. When the dwelling has six or more 2-wire branch circuits, the minimum allowable ampacity of the ungrounded conductors is 50 amperes. II. When the dwelling has a computed load of 10 kVA or more, the minimum allowable ampacity of the ungrounded conductors is 100 amperes.

A) I only
C) II only
B) Both I and II
D) Neither I nor II

98. Which of the following statements about the inspection of electrical work is/are correct? I. Inspectors shall make as many inspections as are necessary to insure compliance with applicable laws. II. An inspector shall issue a certificate of compliance if a completed electrical installation complies with all applicable laws and with the terms of the permit.

A) Both I and II
C) I only
B) Neither I nor II
D) II only

99. A 208-volt, 50 horsepower, three-phase squirrel-cage motor has a full-load current rating of:

A) 143 amperes
C) 162 amperes
B) 130 amperes
D) 195 amperes

100. Which of the following is the maximum allowable rating of a permanently connected appliance where the branch-circuit overcurrent device is used as the appliance disconnecting means?

A) 1/8 horsepower
C) 1/4 horsepower
B) 1/2 horsepower
D) 1 horsepower

Electrician's Exam Answer Sheet

Name ____

Please print (last) (first) (middle)

Address ____

____

Signature ____

1	Ⓐ	Ⓑ	Ⓒ	Ⓓ	26	Ⓐ	Ⓑ	Ⓒ	Ⓓ	51	Ⓐ	Ⓑ	Ⓒ	Ⓓ	76	Ⓐ	Ⓑ	Ⓒ	Ⓓ
2	Ⓐ	Ⓑ	Ⓒ	Ⓓ	27	Ⓐ	Ⓑ	Ⓒ	Ⓓ	52	Ⓐ	Ⓑ	Ⓒ	Ⓓ	77	Ⓐ	Ⓑ	Ⓒ	Ⓓ
3	Ⓐ	Ⓑ	Ⓒ	Ⓓ	28	Ⓐ	Ⓑ	Ⓒ	Ⓓ	53	Ⓐ	Ⓑ	Ⓒ	Ⓓ	78	Ⓐ	Ⓑ	Ⓒ	Ⓓ
4	Ⓐ	Ⓑ	Ⓒ	Ⓓ	29	Ⓐ	Ⓑ	Ⓒ	Ⓓ	54	Ⓐ	Ⓑ	Ⓒ	Ⓓ	79	Ⓐ	Ⓑ	Ⓒ	Ⓓ
5	Ⓐ	Ⓑ	Ⓒ	Ⓓ	30	Ⓐ	Ⓑ	Ⓒ	Ⓓ	55	Ⓐ	Ⓑ	Ⓒ	Ⓓ	80	Ⓐ	Ⓑ	Ⓒ	Ⓓ
6	Ⓐ	Ⓑ	Ⓒ	Ⓓ	31	Ⓐ	Ⓑ	Ⓒ	Ⓓ	56	Ⓐ	Ⓑ	Ⓒ	Ⓓ	81	Ⓐ	Ⓑ	Ⓒ	Ⓓ
7	Ⓐ	Ⓑ	Ⓒ	Ⓓ	32	Ⓐ	Ⓑ	Ⓒ	Ⓓ	57	Ⓐ	Ⓑ	Ⓒ	Ⓓ	82	Ⓐ	Ⓑ	Ⓒ	Ⓓ
8	Ⓐ	Ⓑ	Ⓒ	Ⓓ	33	Ⓐ	Ⓑ	Ⓒ	Ⓓ	58	Ⓐ	Ⓑ	Ⓒ	Ⓓ	83	Ⓐ	Ⓑ	Ⓒ	Ⓓ
9	Ⓐ	Ⓑ	Ⓒ	Ⓓ	34	Ⓐ	Ⓑ	Ⓒ	Ⓓ	59	Ⓐ	Ⓑ	Ⓒ	Ⓓ	84	Ⓐ	Ⓑ	Ⓒ	Ⓓ
10	Ⓐ	Ⓑ	Ⓒ	Ⓓ	35	Ⓐ	Ⓑ	Ⓒ	Ⓓ	60	Ⓐ	Ⓑ	Ⓒ	Ⓓ	85	Ⓐ	Ⓑ	Ⓒ	Ⓓ
11	Ⓐ	Ⓑ	Ⓒ	Ⓓ	36	Ⓐ	Ⓑ	Ⓒ	Ⓓ	61	Ⓐ	Ⓑ	Ⓒ	Ⓓ	86	Ⓐ	Ⓑ	Ⓒ	Ⓓ
12	Ⓐ	Ⓑ	Ⓒ	Ⓓ	37	Ⓐ	Ⓑ	Ⓒ	Ⓓ	62	Ⓐ	Ⓑ	Ⓒ	Ⓓ	87	Ⓐ	Ⓑ	Ⓒ	Ⓓ
13	Ⓐ	Ⓑ	Ⓒ	Ⓓ	38	Ⓐ	Ⓑ	Ⓒ	Ⓓ	63	Ⓐ	Ⓑ	Ⓒ	Ⓓ	88	Ⓐ	Ⓑ	Ⓒ	Ⓓ
14	Ⓐ	Ⓑ	Ⓒ	Ⓓ	39	Ⓐ	Ⓑ	Ⓒ	Ⓓ	64	Ⓐ	Ⓑ	Ⓒ	Ⓓ	89	Ⓐ	Ⓑ	Ⓒ	Ⓓ
15	Ⓐ	Ⓑ	Ⓒ	Ⓓ	40	Ⓐ	Ⓑ	Ⓒ	Ⓓ	65	Ⓐ	Ⓑ	Ⓒ	Ⓓ	90	Ⓐ	Ⓑ	Ⓒ	Ⓓ
16	Ⓐ	Ⓑ	Ⓒ	Ⓓ	41	Ⓐ	Ⓑ	Ⓒ	Ⓓ	66	Ⓐ	Ⓑ	Ⓒ	Ⓓ	91	Ⓐ	Ⓑ	Ⓒ	Ⓓ
17	Ⓐ	Ⓑ	Ⓒ	Ⓓ	42	Ⓐ	Ⓑ	Ⓒ	Ⓓ	67	Ⓐ	Ⓑ	Ⓒ	Ⓓ	92	Ⓐ	Ⓑ	Ⓒ	Ⓓ
18	Ⓐ	Ⓑ	Ⓒ	Ⓓ	43	Ⓐ	Ⓑ	Ⓒ	Ⓓ	68	Ⓐ	Ⓑ	Ⓒ	Ⓓ	93	Ⓐ	Ⓑ	Ⓒ	Ⓓ
19	Ⓐ	Ⓑ	Ⓒ	Ⓓ	44	Ⓐ	Ⓑ	Ⓒ	Ⓓ	69	Ⓐ	Ⓑ	Ⓒ	Ⓓ	94	Ⓐ	Ⓑ	Ⓒ	Ⓓ
20	Ⓐ	Ⓑ	Ⓒ	Ⓓ	45	Ⓐ	Ⓑ	Ⓒ	Ⓓ	70	Ⓐ	Ⓑ	Ⓒ	Ⓓ	95	Ⓐ	Ⓑ	Ⓒ	Ⓓ
21	Ⓐ	Ⓑ	Ⓒ	Ⓓ	46	Ⓐ	Ⓑ	Ⓒ	Ⓓ	71	Ⓐ	Ⓑ	Ⓒ	Ⓓ	96	Ⓐ	Ⓑ	Ⓒ	Ⓓ
22	Ⓐ	Ⓑ	Ⓒ	Ⓓ	47	Ⓐ	Ⓑ	Ⓒ	Ⓓ	72	Ⓐ	Ⓑ	Ⓒ	Ⓓ	97	Ⓐ	Ⓑ	Ⓒ	Ⓓ
23	Ⓐ	Ⓑ	Ⓒ	Ⓓ	48	Ⓐ	Ⓑ	Ⓒ	Ⓓ	73	Ⓐ	Ⓑ	Ⓒ	Ⓓ	98	Ⓐ	Ⓑ	Ⓒ	Ⓓ
24	Ⓐ	Ⓑ	Ⓒ	Ⓓ	49	Ⓐ	Ⓑ	Ⓒ	Ⓓ	74	Ⓐ	Ⓑ	Ⓒ	Ⓓ	99	Ⓐ	Ⓑ	Ⓒ	Ⓓ
25	Ⓐ	Ⓑ	Ⓒ	Ⓓ	50	Ⓐ	Ⓑ	Ⓒ	Ⓓ	75	Ⓐ	Ⓑ	Ⓒ	Ⓓ	100	Ⓐ	Ⓑ	Ⓒ	Ⓓ

Figure 14-9: Answer sheet for use with afternoon exam

Afternoon Exam

Most examining agencies allow approximately one hour for lunch. This is the time to relax and get your body and mind in shape for the afternoon exam.

In the majority of cases, the examining agency will have a place for lunch picked out for those taking the exam to ensure that everyone will be served in time to return for the afternoon exam at the specified time. In some cases, a bus is provided for transporting applicants back and forth if the eating place is not within easy walking distance.

This is not the time for further study. If you aren't prepared for the exam by this time, the few minutes you'll have for further study won't help.

The areas of examination on the afternoon exam will be similar to those on the morning exam:

- The *National Electrical Code*
- General knowledge of electrical practices
- Supporting theoretical knowledge
- Local ordinance installation requirements

However, in many cases, you'll find that questions about the *NEC*, including installation requirements and design calculations, usually comprise from 70% to 80% of the examination. Some examining agencies include test questions on local ordinances and installation requirements in a separate examination.

The questions to follow cover all of these areas and should adequately prepare you for the afternoon electrician's examination.

AFTERNOON EXAM
TIME: 4 HOURS

1. A load is considered to be continuous if the maximum current is expected to continue for:

A) 1 hour
B) 2 hours
C) 3 hours
D) 4 hours

2. The maximum length of Type S cord connecting a built-in dishwasher shall be:

A) 18 inches
B) 24 inches
C) 36 inches
D) 48 inches

3. Type TW copper has an insulation temperature rating of:

A) 60° C
B) 75° C
C) 90° C
D) 110° C

4. Which one of the following is the maximum load current of three (3) No. 12 AWG THWN copper conductors installed in a raceway?

A) 15 amperes
B) 20 amperes
C) 25 amperes
D) 30 amperes

5. Which of the following statements about liquidtight flexible metal conduit is/are correct?
I. Liquidtight flexible metal conduit shall not be used where subject to physical damage.
II. Liquidtight flexible metal conduit shall be used only with listed terminal fittings.

A) I only
B) II only
C) Both I and II
D) Neither I nor II

6. All of the following are standard ampere ratings for inverse time breakers except:

A) 275 amperes
B) 300 amperes
C) 350 amperes
D) 400 amperes

7. A metal outlet device box contains eight (8) No. 12 conductors, no clamps, hickeys, devices or grounding conductors. Which of the following is the minimum allowable box dimension; that is, width and depth?

A) 4 × 1 1/4 inches square

B) 4 × 2 1/8 inches square

C) 4 × 1 1/2 inches square

D) 4 × 2 1/8 inches octagonal or round

8. The conductors of a feeder supplying two (2) continuous duty motors, one rated 7 1/2 horsepower, 240 volts dc, and the other rated 10 horsepower, 240 volts dc shall have a current rating not less than one of the following:

A) 56 amperes

B) 74 amperes

C) 77 amperes

D) 84 amperes

9. An equipment bonding jumper is permitted outside a raceway:

A) When not over 5 feet long

B) When not over 6 feet long

C) When not over 8 feet long

D) When not over 10 feet long

10. The identification of the grounded (neutral) terminal on a polarized receptacle shall be by a metal or metal coating substantially:

A) Yellow in color

B) Green in color

C) Gold in color

D) White in color

11. The maximum allowable ampacity of each of three single insulated 75° C rated No. 6 AWG copper conductors in a raceway is:

A) 55 amperes

B) 65 amperes

C) 70 amperes

D) 80 amperes

12. Which one of the following is the minimum size conductor permitted to be used for the bonding (equipment grounding) conductor in a swimming pool, wading pool, or similar installation?

A) No. 10 stranded copper

B) No. 10 solid copper conductor or No. 12 solid copper-clad aluminum conductor

C) No. 8 solid copper conductor

D) No. 8 stranded copper conductor

13. Which of the following statements is/are correct? I. A swimming pool bonding (SP-SP) electrical contractor must obtain a statewide certificate of competency issued by the Commissioner of Insurance before engaging in business. II. A qualified individual is not required to be regularly on active duty to supervise and direct all work performed under the license on which he is listed.

A) I only

B) Both I and II

C) II only

D) Neither I nor II

14. The total computed load for the electric service shown in Figure 14-10 is 200 amperes for each ungrounded conductor and 60 amperes maximum unbalanced load for the grounded (neutral) conductor. If a THW copper conductor is used, the minimum allowable size of the grounded (neutral) conductor is:

A) No. 6

B) No. 4

C) No. 2/0

D) No. 3/0

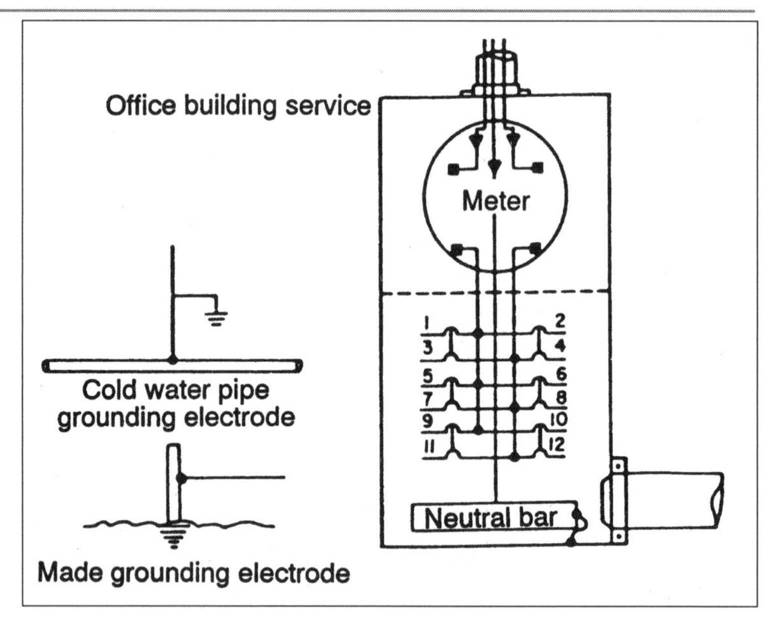

Figure 14-10: Electric service for an office building, with details of the grounding electrodes

15. The minimum allowable ampacity that shall be used to size conductor A in Figure 14-11 is:

A) 260 amperes

B) 361 amperes

C) 551 amperes

D) 451 amperes

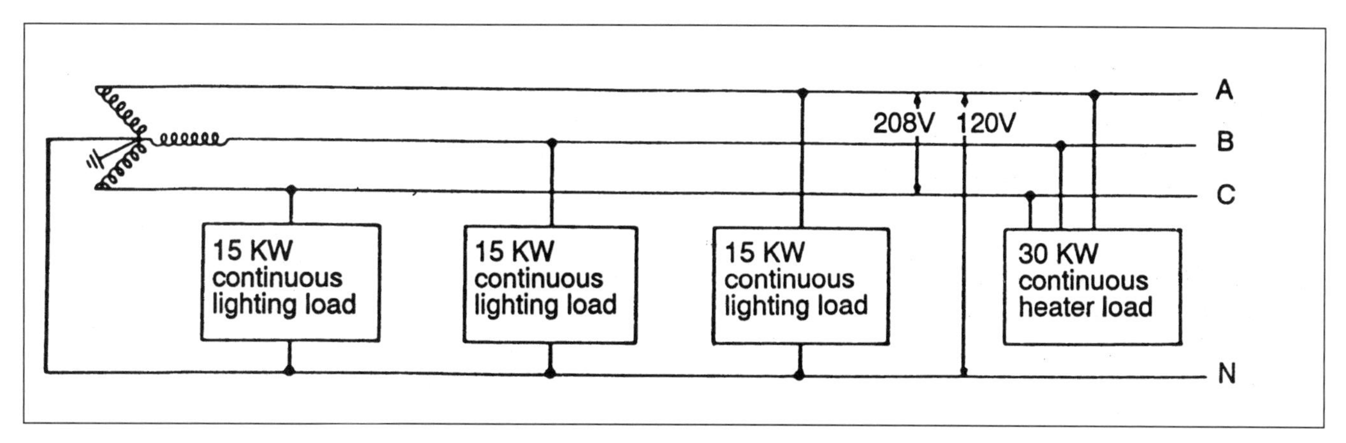

Figure 14-11: Wiring diagram of a three-phase wye-connected electric service

16. A "section" or "cross-section" of an object or a building is what could be seen if the object were:

A) Sliced into two parts with one part removed

B) Sliced into six parts with three parts removed

C) Sawed into four parts

D) Left solid

17. A supplemental drawing used with conventional electrical drawings that gives a complete and more exact description of an item's use is called:

A) Title block

B) Detail drawing

C) Schedule

D) Riser diagram

18. A device that provides protection from the effects of arc faults is known as a(n):

A) GFCI

B) AFGI

C) ACFI

D) AFCI

19. The maximum voltage drop for branch circuits and feeders combined should not exceed:

A) 1% of the circuit voltage

B) 2% of the circuit voltage

C) 3% of the circuit voltage

D) 4% of the circuit voltage

20. The rating of the receptacle providing shore power for a 26-foot boat at a marina shall not be less than:

A) 15 amperes

B) 20 amperes

C) 25 amperes

D) 30 amperes

21. If a multiwire branch circuit supplies line-to-line loads, the branch-circuit protective device must open each:

A) Ungrounded conductor simultaneously

B) All conductors, including the grounded conductor

C) Only the grounded conductor

D) None of the conductors

22. Unless it is bare, an equipment grounding conductor must always be identified by a continuous green color or a continuous green color with:

A) Blue stripes

B) Red stripes

C) Yellow stripes

D) Black stripes

23. Service fuses may be placed on which side of the service disconnecting means?

A) At each outlet

B) Line side

C) Load side

D) Before the last transformer on the system

24. The service disconnecting means can consist of up to what number of switches or circuit breakers without a main disconnect?

A) 2

B) 4

C) 6

D) 8

25. As the area of a wire increases, the dc resistance decreases, but the ac resistance increases as a percentage of the:

A) dc resistance

B) Voltage

C) Impedance

D) Wattage

26. The medium-size lampholder is not permitted on any branch circuit rated over:

A) 15 amperes

B) 20 amperes

C) 30 amperes

D) 40 amperes

27. The *NEC* permits only what type of permanently connected lampholders on 30-ampere circuits other than those used in dwellings?

A) Edison-base

B) Light-duty

C) Mogul or heavy-duty

D) Type S base

28. Where portable appliances are used on a 30-ampere branch circuit, they cannot be rated over:

A) 15 amperes

B) 20 amperes

C) 30 amperes

D) 24 amperes

29. To prevent the use of a portable appliance on a smaller or higher current rated branch circuit, receptacles for portable appliances must be rated for:

A) 15 amperes

B) 20 amperes

C) 30 amperes

D) 40 amperes

30. When both feeders and branch circuits are included, the maximum voltage drop allowed on a system is:

A) 2%

B) 3%

C) 4%

D) 5%

31. Each subpanel or branch-circuit panelboard must be supplied with:

A) A separate feeder

B) At least three feeders

C) Two or more feeders

D) Not less than four feeders

32. A feeder must have an ampacity to provide for:

A) 50% of the building's total electrical load

B) The total building electrical load

C) The maximum anticipated load

D) The minimum calculated load

33. A store or similar occupancy will more than likely operate all its lighting simultaneously; therefore a demand factor of what percent must be used?

A) 90%

B) 100%

C) 125%

D) 150%

34. What size TW copper conductor is required to carry an 18-ampere continuous load?

A) No. 14 AWG
B) No. 10 AWG
C) No. 12 AWG
D) No. 4 AWG

35. What size THHN copper conductor is required to carry a three-phase motor load with a full-load rating of 83 amperes?

A) No. 4 AWG
B) No. 1 AWG
C) No. 2 AWG
D) No. 1/0

36. What size THW copper conductors must be used to feed a noncontinuous, three-phase, 3-wire load of 80 amperes?

A) No. 6 AWG
B) No. 4 AWG
C) No. 2 AWG
D) No. 1 AWG

37. Conductors within electrical nonmetallic tubing may not exceed what voltage?

A) 300 volts
B) 450 volts
C) 500 volts
D) 600 volts

38. Six 230-volt, No. 8 AWG copper conductors with Type TW insulation are to be installed in a single raceway. What is the maximum allowable load current of each conductor?

A) 28 amperes
B) 32 amperes
C) 40 amperes
D) 50 amperes

39. If the service-entrance conductors are size 3/0, the minimum size of a copper grounded neutral conductor is:

A) 3/0
B) 2/0
C) 2
D) 4

40. A 20 horsepower wound-rotor, no code letter, motor is to be installed on a 460-volt, three-phase ac circuit. Disregarding exceptions, the largest nontime-delay fuse to provide short circuit and ground-fault protection for the motor is:

A) 30 amperes

B) 40 amperes

C) 50 amperes

D) 60 amperes

41. On typical wiring diagrams for magnetic motor control starters, overload heaters are shown in series with the:

A) Control circuit supplying the coil of the motor starter

B) Line contacts supplying power to the motor

C) Pilot light that indicates when the motor is stopped

D) Pilot light that indicates when the motor is on

42. A 120-volt lighting fixture has ten 100-watt lamps that are fed through a common fixture wire. The minimum size fixture wire for the one common wire that feeds the entire fixture is:

A) No. 16 AWG

B) No. 14 AWG

C) No. 12 AWG

D) No. 10 AWG

43. Wiring that provides external power to aircraft hangars shall be installed at least how many inches above floor level?

A) 6

B) 12

C) 18

D) 24

44. When service-entrance phase conductors are larger than 1,750 MCM aluminum, the bonding jumper should have an area of not less than what percent of the area of the largest phase conductor?

A) 8.5%

B) 10%

C) 12.5%

D) 20%

45. A five-family apartment building is supplied by a 120/240-volt, 3-wire service. Each apartment has a calculated load of 40 kVA. According to the *NEC* optional multifamily-dwelling calculation, the load of a 3-wire service for this building should not be less than approximately how many kVA?

A) 80 kVA

B) 85 kVA

C) 88 kVA

D) 90 kVA

46. A 200-ampere lighting and appliance branch-circuit panelboard may be protected by two main breakers with a combined rating of:

A) 150 amperes

B) 175 amperes

C) 200 amperes

D) 225 amperes

47. A lighting and appliance panelboard is a panelboard in which more than 10% of its overcurrent devices are rated 30 amperes or less *and* have:

A) Ungrounded connections only

B) GFCIs installed

C) Neutral connections

D) Switched neutral

48. All metal enclosures for service conductors and equipment must be:

A) Grounded

B) Ungrounded

C) Made from 10 gauge steel or above

D) PVC (plastic)

49. On a switchboard with busbars, phase B would have the highest voltage to ground in a:

A) Delta-wye system

B) High-leg delta-connected system

C) Wye-wye system

D) Scott connection

50. The terminal bar in a panelboard is connected to the neutral bar only when the panelboard is used as:

A) Motor control feeders

B) Low-voltage feeders

C) Service equipment

D) Machinery feeders

51. A 1 horsepower electric motor always requires a:

A) Disconnecting means

B) Protection barrier

C) Solid concrete mount

D) Three-wire feeder

52. Electricity is normally generated at power plants between what two voltages?

A) 120 - 460 volts

B) 460 - 600 volts

C) 2,400 - 13,200 volts

D) 600 - 1000 volts

53. What section of the electrical distribution system steps the voltage down to distribution voltages at cities or industrial plants?

A) Transmission lines
B) Pothead
C) Substations
D) Fused cutouts

54. Transformers mounted on the ground are normally referred to as what type of transformers?

A) Submersible
B) Pole-mounted
C) Ground-mounted
D) Padmount

55. Conductors on the high-voltage side of transformers are known as what kind of conductors?

A) Secondary
B) Output
C) Primary
D) Feeders

56. Conversely, conductors on the low-voltage side of transformers are known as what kind of conductors?

A) Secondary
B) Input
C) Primary
D) Feeders

57. What type of device protects connections of underground cable and overhead conductors?

A) Lightning arresters
B) Insulators
C) Potheads
D) Fused cutout

58. What is the name of the device used on electrical distribution systems that protects the circuit against lightning?

A) Insulating bushing
B) Lightning arrester
C) Pothead
D) Fused cutout

59. What name is given to the conductors on the low-voltage side at a substation?

A) Service drop
B) Feeders
C) Branch circuit
D) Service cables

60. Name two types (phases) of current carried by distribution lines.

A) Single- and three-phase
B) Two- and four-phase
C) Five- and six-phase
D) One- and four-phase

61. AFCI's are required on all 15 and 20 ampere circuits that supply:

A) Garages
B) Outdoor receptacles
C) Bedrooms
D) Bathrooms

62. Name the voltage and phase, along with the number of wires, of the most-used electric service for residential occupancies.

A) 240/460 V, three-phase, 3-wire
B) 120/220 V, single-phase, 4-wire
C) 277/480 V, three-phase, 4-wire
D) 120/240 V, single-phase, 3-wire

63. What is the name given to transformers that are installed in underground vaults?

A) Pole-mounted
B) Submersibles
C) Padmount
D) Ground-mounted

64. Transformers mounted on poles are known by what name?

A) Padmount
B) Submersibles
C) Pole-mounted
D) Ground-mounted

65. What type of occupancies normally employ three-phase, 4-wire electric services?

A) Fruit and vegetable stands
B) Residential
C) Commercial
D) Dwellings

66. What name is given to the electrical lines that move high voltage electricity over great distances?

A) Service conductors
B) Transmission lines
C) Branch circuits
D) Feeders

67. What two electrical units are multiplied together to obtain watts?

A) Volts and amps
B) Ohms and watts
C) Watts and volts
D) Ohms and volts

68. How many watts are in one (1) kilowatt?

A) 10 watts
B) 100 watts
C) 1 watt
D) 1,000 watts

69. If one 100-watt lamp is energized for 10 hours, how many kilowatt hours of electricity are used?

A) 10,000
B) 100
C) 1
D) 10

70. Once electricity is generated at a plant, what is done to the voltage before it is transmitted across country?

A) Remains the same
B) Stepped up
C) Stepped down
D) Converted to direct current

71. What flows in conductors to produce electricity?

A) Electrons
B) Ions
C) Molecules
D) Protons

72. In brief, what are electrons thought to consist of?

A) Silver
B) Atomic particles
C) Copper
D) Aluminum

73. What type of cables are usually installed for underground wiring?

A) Uninsulated aluminum
B) MC cable
C) Insulated
D) Aluminum only

74. Below what voltage level is electricity stepped down to at substations?

A) 69,000 volts

B) 150,000 volts

C) 100,000 volts

D) 200,000 volts

75. One section of a substation contains a capacitor bank. Name three other sections.

A) Panelboards, branch circuits, and service conductors

B) Busbars, circuit breakers, and transformers

C) Panelboards, branch circuits, and feeders

D) Conduits, locknuts, and bushings

76. What devices are used in a substation to maintain the system's voltage?

A) Potheads

B) Capacitors

C) Transformers

D) Fused cutouts

77. With what section of the electrical distribution system will most electricians become involved?

A) Secondary

B) Transmission lines

C) Primary

D) Generation system

78. What is one use of a transformer inside a residential occupancy?

A) Reduce transmission voltage to residential use

B) Reduce transmission voltage to substation use

C) Reduce 120 volts for use on a low-voltage signaling system

D) Increase transmission voltage for residential use

79. When alternating current flows through a transformer coil, what type of field is generated around the coil?

A) Alternating magnetic field

B) Magnetic leakage

C) Direct current

D) Non-magnetic leakage

80. When the field in one coil cuts through the turns of a second coil, voltage will be generated in this second coil. What is this induced voltage called?

A) Voltage of reactive induction

B) Voltage of mutual induction

C) Voltage of automatic induction

D) Self induction

81. Name the three parts of a very basic transformer.

A) Core, regulator, and primary winding

B) Delta winding, core, and primary winding

C) Core, primary winding, and secondary winding

D) Inner core, outer core, and secondary winding

82. What would be the voltage on the secondary (output) side of a transformer with a 5:1 winding ratio (primary side has 5 times the number of windings as the secondary side) if the input (primary) voltage is 120 volts?

A) 120 volts

B) 48 volts

C) 240 volts

D) 24 volts

83. What type is the transformer in Question No. 82?

A) Step-up

B) Autotransformer

C) Step-down

D) Pole-mounted transformer

84. All transformers have some minor power losses, but for general wiring for building construction, what percent efficiency may be used in circuit calculations?

A) 50%

B) 75%

C) 60%

D) 100%

85. Name the three basic types of iron-core transformers.

A) Open, closed, and shell

B) Signal, control, and open

C) Dry, liquid-filled, and submersible

D) Open, liquid-filled, and dry

86. What is one effect that magnetic leakage in a transformer causes?

A) Increase in secondary voltage

B) Decrease in secondary voltage

C) Increase in primary voltage

D) Decrease in primary voltage

87. Name two areas where the magnetic path travels in an open-core transformer.

A) North and south poles

B) East and west poles

C) Core and air

D) North pole and air

88. Which one of the three types of iron-core transformers is the most efficient?

A) Shell

B) Dry core

C) Open core

D) Closed core

89. What is the main purpose of transformer taps?

A) To increase efficiency

B) To mount transformer

C) To change the output (secondary) voltage

D) To make connections

90. One of the symptoms of a transformer with an open circuit is:

A) High voltage reading

B) Heating

C) A reading of zero voltage on the secondary

D) Low voltage reading

91. One of the symptoms of a transformer with a partial ground fault is:

A) Cooling off of transformer

B) Overheating of transformer

C) No voltage reading

D) High voltage reading

92. One of the symptoms of a grounded transformer winding is:

A) Overheating and high voltage

B) No voltage

C) Cooling off and high voltage

D) Overheating and low voltage

93. According to Kirchhoff's voltage law, which of the following Ohm's law equations may be used to find the voltage drop across a resistor?

A) Voltage drop = IR

B) Voltage drop = ER

C) Voltage drop = E/R

D) Voltage drop = I/R

94. A 120-volt circuit has an electric heater connected with a current rating of 7.5 amperes. What is the resistance in ohms of the connected pure resistance load?

A) 5 ohms

B) 10 ohms

C) 16 ohms

D) 15 ohms

95. An ohmmeter shows the resistance of a 240-volt electric heater to be 19.5 ohms. What current flows through this heater?

A) 5.9 amperes

B) 20.3 amperes

C) 10.7 amperes

D) 12.3 amperes

96. What current is drawn by a 277-volt fluorescent lamp with 8 ohms reactance?

A) 45

B) 8

C) 34.62

D) 7.4

97. What is the minimum general lighting load permitted by the *NEC* in a 4500 square foot office building?

A) 1125 volt-amperes

B) 15,750 volt-amperes

C) 13,500 volt-amperes

D) 18,000 volt-amperes

98. What is the minimum general lighting load permitted in a 1500 square foot single-family dwelling as specified in Table 220-3(a) of the *NEC*?

A) 1500 volt-amperes

B) 2500 volt-amperes

C) 4500 volt-amperes

D) 3500 volt-amperes

99. A 50 kVA, three-phase, delta-to-wye connected, 480/120-208-volt transformer is used to supply a lighting load. What is the rated line current on the primary side?

A) 75 amperes

B) 17.3 amperes

C) 60.21 amperes

D) 20.70 amperes

100. A 480/277-volt, Y-connected transformer is used to supply a balanced 277-volt, single-phase lighting load of 40,000 watts. What size transformer (kVA) should be used?

A) 10 kVA

B) 20 kVA

C) 40 kVA

D) 50 kVA

Using the Interactive Electrician's Exam Study Center CD

Inside the back cover is a CD with all the questions in the book, including the figures. You can use the CD as either a study tool or a simulated exam. In study mode, you get instant answer feedback with a brief explanation and the corresponding *NEC* reference. In exam mode, you can mark questions to answer later, receive a numerical score, review incorrect answers, retake the test, and print your test results.

To start, insert the **Electrician's Exam Interactive Study Center CD** into your computer's CD-ROM drive. The CD should start the Introduction automatically.

If, after a minute, the CD does not start automatically:

1. Click **Start**, and then click **Run**.
2. Click **Browse** to open the browse window.
3. Navigate to locate your CD-ROM drive, and then double-click to open it.
4. Double-click "**craftsman**" to select the program and return to the Run window.
5. Click **OK** to start the program.

The Introduction lasts a few minutes. To skip the Introduction and go directly to the "*Welcome to the Craftsman Interactive Study Center*" screen, click **Skip**.

On the "*Welcome* . . ." screen, two options appear:

- **Take Test** – Click to begin the study and testing process.
- **Connect to Web** – Click to connect to Craftsman's Web site.

After clicking **Take Test**, the first option is to determine what to study. You can select a specific chapter of the book, or you can select a random set of 25 questions taken from the entire book.

To take the test by Chapter:

1. Click **Chapter** to display page 1 of 2, Chapters 1 – 8.
2. Click the **right ➢ arrow** to display page 2 of 2, Chapters 9 – Final Exam.
3. Click the desired chapter to continue.

- **Back** – Click to return to the prior screen.

To take the test as a random set of 25 questions, click **Random**.

To return to the "*Welcome* . . ." screen, click **Main Menu**.

Once you have selected the topic for study, you are then able to choose to take the test in Study Mode or in Exam Mode.

To take the test in Study Mode:

1. Click **Study Mode** to display the first question.
2. Review the question, and then click your answer.
   - **Correct** answer will yield a green ✓ check mark and a bell will sound.
   - **Incorrect** answer will yield a red **X** and a beep will sound.
   - **Explanation** tab – Click to display a pop-up window with the correct answer, a brief explanation and figure(s). Click **Explanation** again to close the pop-up.

To take the test in Exam Mode:

1. Click **Exam Mode**.
2. Click **Timed Exam** or **Untimed Exam** to display the first question.
3. Review the question, and then click your answer. Move to the next or previous question by using the arrows at the bottom of the screen, or:
   - **Mark** – Click to identify the current question for review later. In Review, marked questions have a red dot; click a question to return to the exam. Click **Mark** again to unmark a question.
   - **Review** – Click anytime during the exam to see your answers, questions you didn't answer and questions you marked for review. In Review, click a number to return to that question. Unanswered questions are ndicated by an orange square.
   - **Finish** – Click to end exam and get your results. If your exam is incomplete, a message will appear with the following options:
     - **Cancel** – Click to return to the exam.
     - **Review Questions** – Click to return to the Review screen.
     - **Results** – Click to see your answers, review incorrect answers, take the exam again, print your results or see the percent of your correct answers. You have three options:
       - **Review Incorrect Answers** – Click to review the question, your answer and the correct answer.
       - **Take Test Again** – Click to return to question #1 of that chapter's exam.
       - **Print Results** – Click to print your answer results.

Appendix I
State Contractor's Examination Offices

State licensing boards are frequently changing their addresses and Web sites and the examining agencies they use. Area codes also change from time to time. While the information provided here was accurate at the time of this printing (January 2005), changes will occur.

If you can't reach an agency at the address given, log on to Craftsman's Web site (www.craftsman-book.com). Scroll to the bottom of the first page and click on the drop down menu at the left to link to Craftsman's *Contractors-License.org*. The information given there is updated routinely every time we learn of an address change. Plus, there are links to many of the state licensing agencies.

Alabama

Electrical Contractors Board
610 S. McDonough Street
Montgomery, AL 36104
(334) 269-9990 Fax (334) 263-6115
www.aecb.state.al.us

Alaska

Division of Occupational Licensing
Construction Contractor Section
P. O. Box 110806
Juneau, AK 99811-0806
(907) 465-8443 Fax (907) 465-2974
www.dced.state.ak.us/occ/

Arizona

Registrar of Contractors
800 W. Washington Street, 6th Floor
Phoenix, AZ 85007
(602) 542-1525
Fax (602) 542-1599
www.rc.state.az.us

Arkansas

Board of Electrical Examiners
10421 W. Markham Street
Little Rock, AR 72205
(501) 682-4549
Fax (501) 682-1765
www.arkansas.gov/labor/divisions

California

Department of Consumer Affairs
Contractors State License Board
9821 Business Park Drive
P. O. Box 26000
Sacramento, CA 95826
(916) 255-3900 (800) 321-2752
Fax (916) 366-9130
www.cslb.ca.gov

Colorado

Department of Regulatory Agencies (DORA)
State Electrical Board
1580 Logan Street, Suite 550
Denver, CO 80203
(303) 894-2300 Fax (303) 894-2310
www.dora.state.co.us

Connecticut

Department of Consumer Protection License Services
165 Capitol Avenue, Room 110
Hartford, CT 06106
(860) 713-6135 Fax (860) 713-7230
www.ct.gov/dcp

Delaware

Division of Professional Regulation
Board of Electrical Examiners
Cannon Building
861 Silver Lake Blvd., Suite 203
Dover, DE 19904
(302) 744-4500 Fax (302) 739-2711
www.professionallicensing.state.de.us/boards/electrician

Florida

Dept. of Business and Professional Regulation – Div. of Professions
Electrical Contractors' Licensing Board
1940 North Monroe Street
Tallahassee, FL 32399-0771
(850) 487-1395 Fax (850) 488-8748
www.state.fl.us/dbpr/pro/elboard/elec_index.shtml

Georgia

State Construction Industry Licensing Board
Division of Electrical Contractors
237 Coliseum Drive
Macon, GA 31217-3858
(478) 207-1416 Fax (478) 207-1425
www.sos.state.ga.us/plb/

Hawaii

Department of Commerce and Consumer Affairs
Professional & Vocational Licensing Electrician & Plumber
P. O. Box 3469
Honolulu, HI 96801
(808) 586-3000 Fax (808) 586-3031
www.hawaii.gov/dcca/pvl

Idaho

Division of Building Safety Electrical Bureau
1090 E. Watertower
Meridian, ID 83642
(208) 334-2183 Fax (208) 855-2165
www.state.id.us/dbs

Illinois

The State of Illinois does not issue State Contractor's Licenses. Licensing is done on a local level. Look in the telephone directory under "Town of," "City of," or "County of."
www.idfpr.com

Indiana

The State of Indiana does not issue State Contractor's Licenses. Licensing is done on a local level. Look in the telephone directory under "Town of," "City of," or "County of."
www.ipla.in.gov

Iowa

The State of Iowa does not issue State Contractor's Licenses. Licensing is done on a local level. Look in the telephone directory under "Town of," "City of," or "County of."
www.iowa.gov

Kansas

The State of Kansas does not issue State Contractor's Licenses. Licensing is done on a local level. Look in the telephone directory under "Town of," "City of," or "County of."
www.accesskansas.org

Kentucky

Office of Housing, Buildings and Construction
Electrical Licensing
101 Sea Hero Road, Suite 100
Frankfort, KY 40601
(502) 573-0382 Fax (502) 573-1598
http://hbc.ppr.ky.gov/electrical_licensing_index.htm

Louisiana

Licensing Board for Contractors
2525 Quail Drive
P. O. Box 14419
Baton Rouge, LA 70898-4419
(225) 765-2301 (800) 256-1392
Fax (225) 765-2690
www.lslbc.louisiana.gov

Maine

Department of Professional and Financial Regulation
Office of Licensing and Registration – Electricians*
35 State House Station
Augusta, ME 04333-0035
(207) 624-8610 Fax (207) 624-8637
www.state.me.us/pfr/olr/categories/cat16.htm
*The State of Maine does not license contractors, only electricians.

Maryland

Division of Occupational and Professional Licensing
State Board of Master Electricians
500 N. Calvert Street, Room 302
Baltimore, MD 21202-3651
(410) 230-6270 Fax (410) 333-6314
www.dllr.state.md.us

Massachusetts

Board of State Examiners of Electricians
Division of Professional Licensure
239 Causeway Street, Suite 500
Boston, MA 02114
(617) 727-9931
www.mass.gov/reg/boards/el

Michigan

Labor & Economic Growth
Construction Codes & Fire Safety
Electrical Division
P. O. Box 30255
Lansing, MI 48909
(517) 241-9320 Fax (517) 241-9308
www.michigan.gov/bccfs

Minnesota

Department of Employment and Economic Development
Electricity Board
1821 University Avenue, S-128
St. Paul, MN 55104-2993
(651) 642-0800 Fax (651) 642-0441
www.electricity.state.mn.us

Mississippi

State Board of Contractors
215 Woodline Drive, Suite B
P. O. Box 320279
Jackson, MI 39232-0279
(601) 354-6161 (800) 880-6161
Fax (601) 354-6715
www.msboc.state.ms.us

Missouri

The State of Missouri does not issue State Contractor's Licenses. Licensing is done on a local level. Look in the telephone directory under "Town of," "City of," or "County of."
www.state.mo.us

Montana

State Electrical Board
301 South Park, 4th Floor
P. O. Box 200513
Helena, MT 59620-0513
(406) 841-2367 Fax (406) 841-2309
www.discoveringmontana.com/dli/ele

Nebraska

State Electrical Division
800 S. 13th Street, Suite 109
P. O. Box 95066
Lincoln, NE 68509-5066
(402) 471-3550 Fax (402) 471-4297
www.electrical.state.ne.us

Nevada

State Contractors Board
2310 Corporate Circle, Suite 200
Henderson, NV 89074
(702) 486-1100 Fax (702) 486-1190
www.nscb.state.nv.us

New Hampshire

Electricians Licensing Board
2 Industrial Park Drive, Building 2
P. O. Box 646
Concord, NH 03302-0646
(603) 271-3748 Fax (603) 271-2257
www.state.nh.us/electrician

New Jersey

Board of Examiners of Electrical Contractors
124 Halsey Street, 6th Floor
P. O. Box 45006
Newark, NJ 07101
(973) 504-6410 Fax (973) 648-3355
www.state.nj.us/lps/ca/nonmed.htm#elec5

New Mexico

Construction Industries Division
2550 Cerrillos Road
P. O. Box 25101
Santa Fe, NM 87504-5101
(505) 476-4700 Fax (505) 476-4685
www.rld.state.nm.us/cid

New York

The State of New York does not issue State Contractor's Licenses. Licensing is done on a local level. Look in the telephone directory under "Town of," "City of," or "County of."
www.dos.state.ny.us

North Carolina

State Board of Examiners of Electrical Contractors
3101 Industrial Drive, Suite 206
P. O. Box 18727
Raleigh, NC 27619-8727
(919) 733-9042 (800) 392-6102
Fax (919) 733-6105
www.ncbeec.org

North Dakota

State Electrical Board
721 Memorial Highway
P. O. Box 857
Bismarck, ND 58502
(701) 328-9522 Fax (701) 328-9524
www.ndseb.com

Ohio

Construction Industry Licensing Board
6606 Tussing Road
P. O. Box 4009
Reynoldsburg, OH 43068-9009
(614) 644-3493 Fax (614) 728-1200
www.com.state.oh.us/dic

Oklahoma

Construction Industries Board
Electrical Licensing Division
2401 N.W. 23rd, Suite 5
Oklahoma City, OK 73107
(405) 271-5217 (877) 484-4424
Fax (405) 271-5254
www.health.state.ok.us/cib

Oregon

Department of Consumer and Business Services
Building Codes Division – Licensing
1535 Edgewater Street, N.W.
P. O. Box 14470
Salem, OR 97309-0404
(503) 378-4133 (503) 373-1268
Fax (503) 378-2322
www.oregonbcd.org

Pennsylvania

The State of Pennsylvania does not issue State Contractor's Licenses. Licensing is done on a local level. Look in the telephone directory under "Town of," "City of," or "County of."
www.dli.state.pa.us

Rhode Island

Department of Labor & Training
Division of Professional Regulation
1511 Pontiac Avenue - Building 70
P. O. Box 20247
Cranston, RI 02920-0943
(401) 462-8571 Fax (401) 462-8528
www.dlt.state.ri.us

South Carolina

Contractors Licensing Board
110 Centerview Drive, Suite 201
P. O. Box 11329
Columbia, SC 29211-1329
(803) 896-4686 Fax (803) 896-4701
www.llr.state.sc.us/pol/contractors

South Dakota

State Electrical Commission
308 S. Pierre Street
Pierre, SD 57501
(605) 773-3573 (800) 233-7765
Fax (605) 773-6213
www.state.sd.us/dcr/electrical

Tennessee

Board of Licensing Contractors
Department of Commerce and Insurance
500 James Robertson Parkway
Suite 110
Nashville, TN 37243-1150
(615) 741-8307 Fax (615) 532-2868
www.state.tn.us/commerce

Texas

Department of Licensing and Regulation – Electricians
920 Colorado
P. O. Box 12157
Austin, TX 78711
(512) 463-6599 (800) 803-9202 (TX)
Fax (512) 475-2871 (no applications)
www.license.state.tx.us

Utah

Division of Occupational and Professional Licensing
160 East 300 South
P. O. Box 146741
Salt Lake City, UT 84114-6741
(801) 530-6628 (866) 275-3675 (UT)
Fax (801) 530-6511
www.dopl.utah.gov

Vermont

Department of Labor and Industry
State Electricians Licensing Board
National Life Building
Drawer 20
Montpelier, VT 05620-3401
(802) 828-2107 Fax (802) 828-2195
www.state.vt.us/labind

Virginia

Board for Contractors
3600 West Broad Street
P. O. Box 11066
Richmond, VA 23230-1066
(804) 367-8511 Fax (804) 367-2474
www.dpor.virginia.gov

Washington

Department of Labor and Industries
Electrical Program
7273 Linderson Way, S.W.
Tumwater, WA 98501
P. O. Box 44460
Olympia, WA 98504-4460
(360) 902-5269 Fax (360) 902-5296
www.lni.wa.gov

West Virginia

State Fire Marshal
Licensing Division
1207 Quarrier Street, 2nd Floor
Charleston, WV 25301
(304) 558-2191 Fax (304) 558-2537
www.wvfiremarshal.org

Wisconsin

Safety and Building Division
201 West Washington Avenue
P. O. Box 7082
Madison, WI 53707-7082
(608) 261-8500 Fax (608) 267-0592
www.commerce.wi.gov/sb/sb-formcredapplist.html

Wyoming

State Fire Marshal's Office
Department of Fire Prevention and Electrical Safety
Herschler 1 West
Cheyenne, WY 82002
(307) 777-7288 Fax (307) 777-7119
http://wyofire.state.wy.us

Appendix II
Answers to Final Exam Questions

Morning Exam

1. C
2. D
3. A
4. C
5. A
6. B
7. A
8. D
9. B
10. A
11. C
12. C
13. C
14. B
15. D
16. D
17. B
18. C
19. B
20. C
21. C
22. B
23. D
24. C
25. B
26. C
27. B
28. B
29. B
30. C
31. A
32. B
33. A
34. B
35. D
36. D
37. C
38. A
39. B
40. B
41. C
42. C
43. A
44. D
45. B
46. B
47. C
48. D
49. C
50. B
51. B
52. B
53. A
54. B
55. D
56. A
57. A
58. C
59. C
60. C
61. A
62. B
63. A
64. D
65. D
66. A
67. C
68. B
69. D
70. C
71. B
72. D
73. A
74. A
75. D
76. D
77. C
78. A
79. C
80. B
81. C
82. A
83. A
84. D
85. B
86. B
87. C
88. C
89. B
90. B
91. B
92. D
93. B
94. B
95. C
96. D
97. C
98. A
99. A
100. A

Afternoon Exam

1. C
2. D
3. A
4. B
5. C
6. A
7. A
8. C
9. B
10. D
11. B
12. C
13. D
14. A
15. A
16. A
17. B
18. D
19. C
20. D
21. A
22. C
23. C
24. C
25. C
26. B
27. C
28. D
29. B
30. D
31. A
32. B
33. B
34. B
35. A
36. B
37. D
38. B
39. D
40. B
41. B
42. B
43. C
44. C
45. D
46. C
47. C
48. A
49. B
50. C
51. A
52. C
53. C
54. D
55. C
56. A
57. C
58. B
59. B
60. A
61. C
62. D
63. B
64. C
65. C
66. B
67. A
68. D
69. C
70. B
71. A
72. B
73. C
74. A
75. B
76. B
77. A
78. C
79. A
80. B
81. C
82. D
83. C
84. D
85. A
86. B
87. C
88. A
89. C
90. C
91. B
92. D
93. A
94. C
95. D
96. C
97. B
98. C
99. C
100. C

Index

D

E

F

G

H

I

J

K

L

M

N

O

P

Q

R

Practical References for Builders

National Repair & Remodeling Estimator

The complete pricing guide for dwelling reconstruction costs. Reliable, specific data you can apply on every repair and remodeling job. Up-to-date material costs and labor figures based on thousands of jobs across the country. Provides recommended crew sizes; average production rates; exact material, equipment, and labor costs; a total unit cost and a total price including overhead and profit. Separate listings for high- and low-volume builders, so prices shown are specific for any size business. Estimating tips specific to repair and remodeling work to make your bids complete, realistic, and profitable. Includes a CD-ROM with an electronic version of the book with *National Estimator*, a stand-alone *Windows*™ estimating program, plus an interactive multimedia video that shows how to use the disk to compile construction cost estimates. **296 pages, 8½ x 11, $53.50. Revised annually**

Craftsman's Construction Installation Encyclopedia

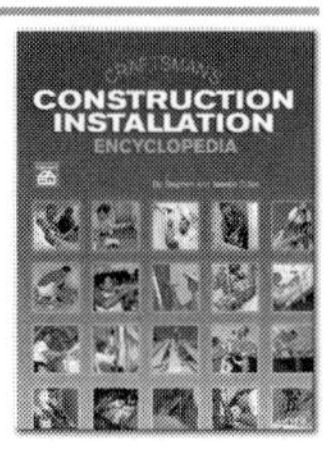

Step-by-step installation instructions for just about any residential construction, remodeling or repair task, arranged alphabetically, from *Acoustic tile* to *Wood flooring*. Includes hundreds of illustrations that show how to build, install, or remodel each part of the job, as well as manhour tables for each work item so you can estimate and bid with confidence. Also includes a CD-ROM with all the material in the book, handy look-up features, and the ability to capture and print out for your crew the instructions and diagrams for any job. **792 pages, 8½ x 11, $65.00**

Contractor's Survival Manual Revised

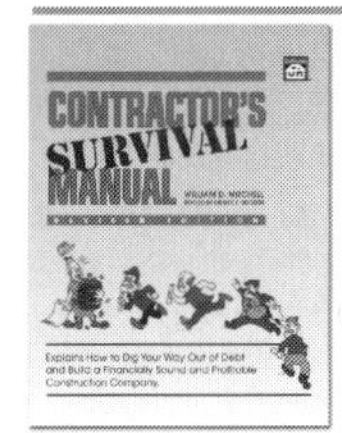

What does it really take to survive hard times in the construction industry? And how do you take full advantage of the profit cycle in good economic times? This practical manual will suggest unique ways to overcome your most persistent problems. During a debt crisis, what do you do when you can't pay the bills? And where can you find money and buy time, transfer debt, handle creditors, choose assets to liquidate, and set payment priorities? If you don't know the answers to these questions, this book is for you. It also tells you cash float techniques, alternatives to bankruptcy, how to deal with lawsuits, judgments and liens and — most important — how to lay the foundation for recovery. Building sales and profits using other people's cash, the ins and outs of limited partnerships, increasing financial reserves, accurate estimating, calculating overhead, contingency and profit margins — all of these are covered. Completely updated, with major new sections on using personal computers for bookkeeping, estimating and scheduling, and Web addresses that have additional information. You won't find conventional advice in this book. Instead, expect to learn what's really needed to survive, stabilize and thrive as a construction contractor. **344 pages, 8½ x 11, $38.00**

National Electrical Estimator

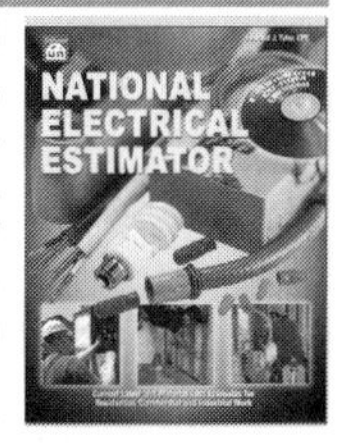

This year's prices for installation of all common electrical work: conduit, wire, boxes, fixtures, switches, outlets, loadcenters, panelboards, raceway, duct, signal systems, and more. Provides material costs, manhours per unit, and total installed cost. Explains what you should know to estimate each part of an electrical system. Includes a CD-ROM with an electronic version of the book with *National Estimator*, a stand-alone *Windows*™ estimating program, plus an interactive multimedia video that shows how to use the disk to compile construction cost estimates. **552 pages, 8½ x 11, $52.75. Revised annually**

Low Voltage Wiring

Take advantage of the boom in security and low voltage systems. Here you'll find answers to designing, installing, maintaining, and troubleshooting security and fire alarm systems in residential and commercial buildings. Explains how to understand blueprints for low voltage systems, what the code requirements are, and how to maximize your profit as a low voltage contractor. Includes a CD-ROM with the checklists and forms to help ensure your success as a low voltage contractor. **408 pages, 8 x 10, $39.95**

Contractor's Guide to QuickBooks Pro 2007

This user-friendly manual walks you through QuickBooks Pro's detailed setup procedure and explains step-by-step how to create a first-rate accounting system. You'll learn in days, rather than weeks, how to use QuickBooks Pro to get your contracting business organized, with simple, fast accounting procedures. On the CD included with the book you'll find a QuickBooks Pro file for a construction company (You open it, enter your own company's data, and add info on your suppliers and subs.) You also get a complete estimating program, including a database, and a job costing program that lets you export your estimates to QuickBooks Pro. It even includes many useful construction forms to use in your business. **344 pages, 8½ x 11, $53.00**

Also available: Contractor's Guide to QuickBooks Pro 2001, $45.25
Contractor's Guide to QuickBooks Pro 2002, $46.50
Contractor's Guide to QuickBooks Pro 2003, $47.75
Contractor's Guide to QuickBooks Pro 2004, $48.50
Contractor's Guide to QuickBooks Pro 2005, $49.75

CD Estimator

If your computer has *Windows*™ and a CD-ROM drive, CD Estimator puts at your fingertips over 130,000 construction costs for new construction, remodeling, renovation & insurance repair, home improvement, framing and finish carpentry, electrical, concrete & masonry, painting, and plumbing & HVAC. Monthly cost updates are available at no charge on the Internet. You'll also have the *National Estimator* program - a stand-alone estimating program for *Windows*™ that *Remodeling* magazine called a "computer wiz," and *Job Cost Wizard*, a program that lets you export your estimates to QuickBooks Pro for actual job costing. A 60-minute interactive video teaches you how to use this CD-ROM to estimate construction costs. And to top it off, to help you create professional-looking estimates, the disk includes over 40 construction estimating and bidding forms in a format that's perfect for nearly any *Windows* word processing or spreadsheet program. **CD Estimator is $78.50**

Building Contractor's Exam Preparation Guide

Passing today's contractor's exams can be a major task. This book shows you how to study, how questions are likely to be worded, and the kinds of choices usually given for answers. Includes sample questions from actual state, county, and city examinations, plus a sample exam to practice on. This book isn't a substitute for the study material that your testing board recommends, but it will help prepare you for the types of questions — and their correct answers — that are likely to appear on the actual exam. Knowing how to answer these questions, as well as what to expect from the exam, can greatly increase your chances of passing.
320 pages, 8½ x 11, $35.00

Commercial Electrical Wiring

Make the transition from residential to commercial electrical work. Here are wiring methods, spec reading tips, load calculations and everything you need for making the transition to commercial work: commercial construction documents, load calculations, electric services, transformers, overcurrent protection, wiring methods, raceway, boxes and fittings, wiring devices, conductors, electric motors, relays and motor controllers, special occupancies, and safety requirements. This book is written to help any electrician break into the lucrative field of commercial electrical work. Updated to the 1999 *NEC*.
320 pages, 8½ x 11, $36.50

Construction Forms & Contracts

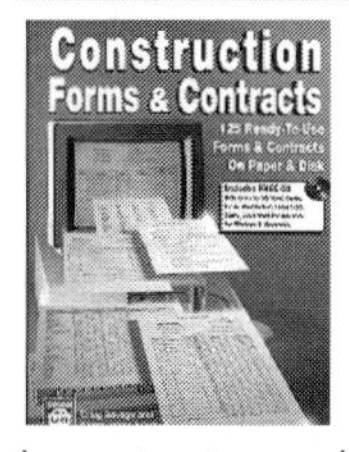

125 forms you can copy and use — or load into your computer (from the FREE disk enclosed). Then you can customize the forms to fit your company, fill them out, and print. Loads into *Word* for *Windows*, *Lotus 1-2-3*, *WordPerfect*, *Works*, or *Excel* programs. You'll find forms covering accounting, estimating, fieldwork, contracts, and general office. Each form comes with complete instructions on when to use it and how to fill it out. These forms were designed, tested and used by contractors, and will help keep your business organized, profitable and out of legal, accounting and collection troubles. Includes a CD-ROM for *Windows*™ and Mac. **432 pages, 8½ x 11, $41.75**